Technology and Isolation

By reconsidering the theme of isolation in the philosophy of technology, and by drawing upon recent developments in social ontology, Lawson provides an account of technology that will be of interest and value to those working in a variety of different fields. *Technology and Isolation* includes chapters on the philosophy, history, sociology and economics of technology and contributes to such diverse topics as the historical emergence of the term 'technology', the sociality of technology, the role of technology in social acceleration, the relationship between Marx and Heidegger and the relationship between technology and those with autism. The central contribution of the book is to provide a new ontology of technology. In so doing, Lawson argues that much of the distinct character of technology can be explained or understood in terms of the dynamic that emerges from technology's peculiar constitutional mix of isolatable and non-isolatable components.

CLIVE LAWSON is currently Director of Studies in Economics and Senior College Lecturer at Girton College, Cambridge, as well as Assistant Director of Studies at Gonville and Caius College. He is an editor of the *Cambridge Journal of Economics* and a founder member of the Cambridge Social Ontology Group. Lawson has published in economics, geography, psychology, sociology, philosophy and environmental economics.

Technology and Isolation

CLIVE LAWSON
University of Cambridge

CAMBRIDGE
UNIVERSITY PRESS

CAMBRIDGE
UNIVERSITY PRESS

University Printing House, Cambridge CB2 8BS, United Kingdom

One Liberty Plaza, 20th Floor, New York, NY 10006, USA

477 Williamstown Road, Port Melbourne, VIC 3207, Australia

314-321, 3rd Floor, Plot 3, Splendor Forum, Jasola District Centre, New Delhi - 110025, India

79 Anson Road, #06-04/06, Singapore 079906

Cambridge University Press is part of the University of Cambridge.

It furthers the University's mission by disseminating knowledge in the pursuit of education, learning and research at the highest international levels of excellence.

www.cambridge.org
Information on this title: www.cambridge.org/9781316632352
DOI: 10.1017/9781316848319

First published 2017
First paperback edition 2018

A catalogue record for this publication is available from the British Library

ISBN 978-1-107-18083-3 Hardback
ISBN 978-1-316-63235-2 Paperback

Technology and Isolation

CLIVE LAWSON
University of Cambridge

CAMBRIDGE
UNIVERSITY PRESS

University Printing House, Cambridge CB2 8BS, United Kingdom

One Liberty Plaza, 20th Floor, New York, NY 10006, USA

477 Williamstown Road, Port Melbourne, VIC 3207, Australia

314-321, 3rd Floor, Plot 3, Splendor Forum, Jasola District Centre, New Delhi - 110025, India

79 Anson Road, #06-04/06, Singapore 079906

Cambridge University Press is part of the University of Cambridge.

It furthers the University's mission by disseminating knowledge in the pursuit of education, learning and research at the highest international levels of excellence.

www.cambridge.org
Information on this title: www.cambridge.org/9781316632352
DOI: 10.1017/9781316848319

First published 2017
First paperback edition 2018

A catalogue record for this publication is available from the British Library

ISBN 978-1-107-18083-3 Hardback
ISBN 978-1-316-63235-2 Paperback

For Lucy, Jesse and Callum

Contents

Preface

It has often been suggested that technology, whatever its benefits, comes at the expense of more isolated and impoverished human lives. This has been a recurrent theme in the philosophy of technology, especially that influenced by Heidegger, where modernity reduces everything – including us – to resources ready for optimisation and control. But the idea will also be familiar to readers of dystopian science fiction, in which technologically sophisticated societies rarely contain any recognisable or meaningful form of human community. More technology, it would seem, leads to more isolation, be it isolation of humans from nature or from each other.

In recent times, however, such ideas have become less prominent. One important reason for this is that some of the most dominant technologies of our time, such as the internet, facilitate a connectivity between people that is unlike anything we have ever known. How can the general tendency of adopting more technology result in greater isolation? One of the main motivations of this book is the intuition that, whilst it is impossible to make such simple pronouncements as 'more technology means more isolation', there are some good reasons why the theme of isolation recurs throughout discussions of technology. Although in need of substantial modification, there is much in these older debates about isolation and separation that are still of significance to current (increasingly technology-reliant) societies, despite the fact that we can so easily Skype our family or play music with strangers on other continents over the internet.

To recover more interesting conceptions of isolation and the different senses in which these have featured in older literatures, I argue, requires a return to ontology. To suggest a turn to ontology is not likely to be treated with the kind of immediate disdain it would have provoked even a few years ago. Indeed, it is almost possible to say that first critical realism and then more recently actor network theory and speculative realism, have made ontology, if not fashionable, then certainly 'acceptable' in many quarters. However, it is also fair to say that these developments in ontology, for different reasons, have not really made much of a contribution to understanding the nature of technology, even though there seems to be great scope for doing so. One of the main concerns of this book is to develop an ontology of technology that draws upon these recent developments in social ontology.

It seems hardly contentious to suggest that we live in a world of *things*. Indeed, the idea that the world is full of things with different causal powers, affordances, organisational properties, etc., seems essential to everyone's ability to get by in the world. However, it is equally uncontentious, I think, to suggest that things operate within and on the basis of being components in different systems. Not all things have similar properties or ways of acting, and many of these differences depend upon differences in the way they are organised. Given this, a central question is the extent to which some things can be understood to operate in relative isolation from other things. Although philosophy and the social sciences are replete with attempts to provide general answers to this question, ranging from reductionist individualisms to emergentist holisms, there seems little doubt that in practice some things are more isolatable than others; some things can operate and be the kinds of things that they are, relatively independently of their relations to other things, others cannot. Moreover, the extent to which things can be understood in isolation is for the most part a matter of empirical discovery. There is not much that can be said a priori.

A central assumption of this book is that the social and non-social worlds are rather different from each other when it comes to matters of isolatability, and that these differences underlie the various ways of being and dynamics of different phenomena, as well as setting constraints on the methods that can be used to understand them. Moreover, issues of isolatability, though neglected, are of particular importance when it comes to the study of things that incorporate aspects of both the social and the non-social in a fundamental way, such as technology.

Although the term 'technology' is one we all capably use on a regular basis, there are actually surprisingly few attempts, across all social sciences and social theory, to pin down exactly what we all mean by the term. A clear example of this is to be found in economics, where, amongst the mainstream at least, there is little interest in developing ideas about the nature of technology. Effectively, anything that changes the relationship between inputs and outputs of some production process is referred to as technology; once we know (which of course we never do) the shape of the functional relationship between inputs and outputs, then no more knowledge of technology is required.

This book arose out of an attempt to fill this rather obvious gap by drawing upon the social ontology with which I was most familiar to try to spell out exactly what we mean by technology. One 'quick' paper divided into three or four relatively unsatisfying journal articles, and it became obvious that a book was required to make even a stab at the project I had set myself. However, the main argument that I want to make is very simple. Technology is made up of both social and non-social elements. These elements, in turn, are susceptible to different amounts of isolatability. Whilst the boundaries between the social and the non-social and the isolatable and the non-isolatable are often porous and dynamic, much that we know of the character of natural and social science, such as the status of controlled experiment in each, suggests that there are huge differences in isolatability in each domain. I will argue that much of the distinct character of technology, and our relationship to it, as well as many of the significant contributions to the study of technology, can be explained or understood in terms of the particular dynamic that emerges from technology's constitutional mix of isolatable and non-isolatable components.

I believe this dynamic explains all kinds of phenomena from economic growth, to the special relationship that those with autism tend to have with technology, to many of the criticisms levelled at current capitalism or modernity. However, to get to these arguments requires some prior setting up of the basic account of technology I want to defend. I must apologise to those readers who will find the preliminary chapters too slow and/or repetitive. But many will find, depending upon their background, that one or more of the early, introductory chapters covers familiar material that can be skipped without losing the argument the book is trying to make.

In short, this book is motivated by a concern with ideas that have currently fallen out of favour but which I believe are as important as they have ever been. It is not a book that attempts to establish whether the net effect of more technology means less community or more isolated people. Neither is it concerned with which technologies tend to connect us and which do not (although the ontology developed in this book should be useful in answering either question). Rather, it is a book about the nature of technology more generally. By focusing upon the way that different ideas of isolation weave in and out of a variety of historical understandings and debates about technology, this book attempts to ground and give meaning to a particular, novel account of technology that is itself set in terms of a particular approach to ontology.

Acknowledgements

This book has been gestating for some time. Consequently, there are a lot of people to thank. The idea for the project first arose whilst on leave visiting the University of Western Sidney and Brian Pinkstone in particular. The main writing was done over two further sabbatical leaves in Vancouver, one visiting the University of British Columbia, the other visiting Simon Fraser University. For the first visit I am very grateful for the welcome and support from Margaret Schabas in particular. On the second visit I am indebted to Andrew Feenberg who gave very unselfishly of his time and stopped me from wasting my time reinventing the wheel. More recently, most of the main writing has taken place in a small cottage in the Breton Beacons – many thanks to Leslie Turano and Chris Taylor.

The main intellectual debt lies with members of the Cambridge Social Ontology Group. Weekly meetings, thrashing out ideas about everything from markets and commodities to the ontology of traffic lights and photocopiers, have provided the main ideational resources for the book. I am especially grateful to Bahar Araz, Phil Faulkner, Tony Lawson, Helen Mussel, Stephen Pratten and Mary Wrenn for giving up their time not only to read earlier drafts of these chapters but to then patiently go through it all with me. In particular I would like to thank Tony Lawson, Stephen Pratten and Phil Faulkner for extensive comments on earlier versions of most, if not all, of these chapters. I would also like to thank Mark Burgess, Lucy Delap, John Lawson and Altuğ Yalçıntaş for comments on specific chapters.

Lastly, I want to thank my family – Lucy, Jesse and Callum – who were so wonderfully supportive and understanding over what must have seemed (and in Callum's case just about was) a lifetime.

1 | *Technology Questions*

It is often suggested that we live in a technological age. Although it is rarely made clear exactly what this statement means, or why or in what ways previous ages are thought not to be technological, most of us seem to agree that technology plays an important role in our lives. We may also agree that this role is becoming increasingly important. In all kinds of daily activities such as buying a bus ticket, guessing how the weather will change, listening to music, paying for shopping at a supermarket, archiving family photos or borrowing a book from the library, we all experience a constant prodding to our routine or 'normal' ways of doing things that can be traced to some or other new technological development. For the most part, moreover, such developments come into being in ways, and for reasons, that lie outside of our control. In other words, we all, through our everyday activities, experience technology's power as an external agent of change.

This experience suggests a range of important questions. To what extent is it possible or desirable to influence the introduction of new technology? To what extent do different technologies determine or constrain the kinds of social changes that follow or accommodate them? Do societies have broad trends or characteristics that are related to the amount or form of technology that have emerged within them – for example, can it be said that people are more or less connected to each other in virtue of the technology they use? Does technology bring with it opportunities for a better life or tend to smuggle in unnecessary problems? Does the form or speed of change of different technologies matter? Is technology always neutral, only taking on good or bad features in some particular context of use? Is it even possible or meaningful to talk in general about 'technology' at all?

Such questions will be familiar to many if not most of us. They have been the lifeblood of science fiction since the beginning of the genre. More formally, or academically, such questions have occupied a wide variety of social theorists since at least the time of the ancient Greeks.

Recently, contributions concerned with such questions have been organised together as constituents of a 'philosophy of technology'.
However, it is fair to say that there remains widespread ambivalence
towards this newly established discipline. An example of such ambivalence is the fact that as interest in this new discipline is growing, interest
in its core questions, such as those above, appears to be waning; in the
midst of 'the technological age', those questions that we might call the
'classic' technology questions are receiving relatively less attention.
Indeed, such questions seem to be currently very unfashionable.

I want to argue, however, that contrary to recent trends, there is
actually much to be gained from systematically pursuing just the kinds
of questions noted above. Not only is it the case that these questions are
(still) in need of answering, but I want to argue that now is a very good
time to address them from a particular perspective. Specifically, this
book approaches such issues from a perspective that owes much to
recent developments in social theorising, in particular in social ontology, which have as yet received very little interest from theorists of
technology. To be clear, neither have social ontologists shown much
interest in technology, nor have philosophers of technology shown
much interest in the kinds of ontological developments I have in
mind. An important motivation of this book is the desire to draw out
connections between these two sets of contributions and initiate
a dialogue between them.

The rest of this chapter is given over to providing an introduction to
the relatively new discipline of the philosophy of technology, and to
explaining the apparently contradictory fact that as the philosophy of
technology has received increasing interest, there has been an identifi-
able ambivalence towards its classic problems and questions. Such an
undertaking is also strategically helpful, in that it helps to contextualise
the arguments that follow in the rest of this book. The starting point,
however, is to give a brief overview of the subject matter to which the
label 'philosophy of technology' is usually understood to refer, which is
the focus of the following section.

On a note of qualification, however, I should point out that I am not
at this stage advancing my own conception or definition of technology.
Rather the point here is to provide the reader, especially if unfamiliar
with the philosophy of technology, with a feel for the kinds of problems
and issues that have concerned those usually understood to be contributing to the philosophy of technology. Such a strategy may give rise to

some apparently contradictory conceptions of 'technology', but I shall delay stating exactly what I mean by the term until after the historical discussion of the emergence of the term 'technology' in the following chapter. The reason for doing this is to explicitly incorporate existing meanings and understandings of technology where possible and helpful. As such, the themes discussed in the remainder of this and much of the next chapter serve to introduce the main ideas that a convincing account of technology should be able to accommodate.

The Philosophy of Technology

At risk of severe oversimplification, two broad themes have dominated the philosophy of technology. The first might be briefly described as the moral or ethical evaluation of technology (or more narrowly, the relation of technology to 'the good life'), whilst the second focuses upon the ways in which our lives are constrained, transformed or controlled by technology (especially as formalised in theories of technological determinism or technological autonomy). Although clearly connected, these two broad themes are initially discussed separately, in turn.

It is fair to say that the recent spate of readers on and companions to the philosophy of technology reveal an intellectual landscape in which evaluative attitudes to technology swing back and forth over time.[1] Typically, a story is told of an initial scepticism towards technological ideas that is reversed by an enlightenment optimism, then replaced by a romantic ambivalence or 'unease', which is itself eventually replaced by some kind of neutrality view of technology (see for example Mitcham, 1994).[2]

The philosophy of technology is usually presented as having its origins in ancient Greece, in the ideas of Socrates, Plato and Aristotle. Once more, two broad themes tend to be highlighted. The first revolves around attempts to create (or defend beliefs about) hierarchies of types of knowledge and learning. For example, there is a distinction made

---

[1] See for example Scharff and Dusek (2003), Dusek (2006), Kaplan (1964), Meijers (2009).

[2] Such retrospective demarcation of the domain of philosophy of technology must, of course, be treated quite cautiously as a summary of ideas about technology, given that it is not clear that the term 'technology' is used in exactly the same manner throughout the contributions highlighted.

between craft, political and philosophical knowledge. Put simply, whilst the craft knowledge of those such as farmers and artisans, is more 'true' and 'honest' than knowledge of a political kind, it still falls some way short of the 'wisdom' available to philosophers (knowledge of the good life arrived at by those who love knowledge). At the heart of these distinctions is the belief that general knowledge is of a higher order than particular or specialised (including technical) knowledge (Dusek, 2006).

This distinction between craft and other kinds of knowledge, relates directly to the second theme, often presented in terms of some kind of scepticism. Greek philosophers tended to believe that although technical knowledge is a necessary part of life, it is in some sense also bad or dangerous. These ideas are evident in a range of stories and myths such as Daedalus and Icarus, the Tower of Babel, and Prometheus. Each of these stories embodies the idea that a preoccupation with technological matters involves a turning away from something good (usually faith in God or nature) and an undermining of individual striving for excellence.[3]

This largely negative or suspicious orientation towards technology is not substantially revised until the writings of Francis Bacon (1561–1626). For Bacon, in contrast to the Greeks, technical knowledge is superior to all other kinds of knowledge and technological artefacts are thought to be inherently good in nature, with any possible dangers being viewed as accidental, or a sign of 'misuse'. Moreover, not only does a turn to technology *not* involve a turn away from God, as seemed to be the case for the ancient Greeks, but according to Bacon a knowledge of nature and its technological uses can be employed by humans to achieve a 'purity of mind and behaviour lost in the "Fall" from the Garden of Eden'.[4] For Bacon, God had given a clear mandate to pursue technology in order to relieve human suffering. Moreover, because humans are created in God's image, it is inevitably in the nature of humans to create and innovate.

---

[3] '[Technology] according to these myths, although to some extent required by humanity and thus on occasion a cause for legitimate celebration easily turn against the human by severing him or her from some larger reality. This severing manifesting in a failure of faith or shift of the will, a refusal to rely on or trust God or the gods' (Mitcham, 1994).

[4] See for example Bacon (1909).

Traces of Bacon's irrepressibly progressive conception of technology can be found in the work of a range of other well-known thinkers.[5] And even as the Renaissance dawned, with its recasting of questions of theological obligation, the belief that humans are effectively unable to live without technology of some form remained. At the same time the pursuit of technology was thought to have positive effects not only on the morals, but the well-being of humans.[6] In this context, the status of those thought 'expert' in technological matters began to increase, thus initiating debates that would later revolve around conceptions of technocracy, early versions of which can be found in the works of Auguste Comte (1798–1857) and Saint Simon (1760–1825).

This new optimism in the role that could be played by technology emerged at the same time as the unprecedented power that was unleashed in the Industrial Revolution. However, this wealth of power seemed to generate countervailing attitudes towards technology in very general terms. On the one hand, there was widespread awe of the possibilities being opened up by the wealth of inventions and innovations of the period. On the other hand, there emerged a distrust of the actual results of such developments. In fact, the real-life consequences of that revolution prompted a range of contributions that were more critical of technological advance. Within such writings, Bacon's ideas were now held up as the main foil against which criticisms of the enlightenment were made. Perhaps the most prominent of these critics was Jean-Jacques Rousseau (1712–1778). Whereas for the Greeks technology was, essentially, bad but necessary, Rousseau attacked what he saw as a complacent progressivism. For Rousseau, the progress of the sciences would lead to decline and decadence, especially destroying the 'virtue and vigour' of the barbarian nations (see for example Rousseau, 1992). Rousseau's anti-Baconianism played an important role in the Romantic critique of the Industrial Revolution more generally and, especially in Germany and England, it played a part in the general sea change of ideas in which, for instance, an organic conception of reality emerged to challenge Newtonian mechanics, and in which logic and reason were counterposed to imagination and feeling.

[5] One notable, relatively subtle, form is taken by Kant, see especially (Kant, 1784).
[6] This, for example, is the position adopted by Hume (Hume, Green and Grose, 2001).

Many of the contributions that followed, however, fell in between this Baconian belief in technological progress and Rousseau's romantic aversion to it. For example, Karl Marx occupied a complex position which drew upon both developments. As I make clear in Chapter 11, although Marx was very critical of the simple Baconian utopias of those such as Saint Simon and Comte (Marx shared the romantic distrust of the employment of technology in the short run, especially under the conditions in which capitalism was in full swing), he was clearly optimistic about the long-run possibilities of technology's benefits for mankind.

More recently, a fourth position has emerged in which there is a tendency, especially evident in the recent constructivist literature, to criticise all positions that imply either an essentially (or generally) good or bad character of technological knowledge or artefacts. Any evaluative statements about technology are understood to be misguided or simply false. Technology, if the term has any meaning at all at such a general level of analysis, can only refer to neutral means to some end or other and it is the ends that might be evaluated in some limited fashion.

Before focusing upon this constructivist position, however, I should say a little more about the position against which it is so often presented as a reaction – technological determinism. As noted above, technological determinism, along with the thesis of technological autonomy, features centrally in the second core theme of the philosophy of technology, which is centrally concerned with the idea that technology constitutes some kind of power or force that is largely independent of the human will.

Jacques Ellul is perhaps the most prominent theorist of the autonomy of technology (Ellul, 1964, 1980). At the heart of Ellul's contribution is the idea that human control of technology is not as capable (or real) as we would like to think. Ellul often uses the term 'technique' to refer to the way that much of our daily activity is brought into conformity with strictly laid-out rules and regulations that increasingly reduce reason to the instrumental. In the process of producing this conformity, the ways in which we are thought to control technology become, rather, a response to the requirements of technology itself. This is especially true, Ellul argues, for those authorities or organisations most thought to be in a position to control and modify technology to our purposes.

Ellul spends a good deal of time explaining why we are blind to such processes. Experts overestimate their own skills, scientists and engineers display embarrassing naiveté with respect to the social implications of different technologies, everyday users of technology increasingly give up any attempts to control technological phenomena, leaving such matters to the experts. Moreover, Ellul argues, the technological system as a whole 'entrances' us all (technologists, politicians, consumers). Advertising changes our desires and the endlessly creative force of technological values displace traditional morality (Ellul, 1964). Aspects of Ellul's account are often mixed with the contributions of others to provide a more sophisticated picture of technology's properties. For example Winner's idea of 'technological somnambulism' mixes the properties of technology with human complacency and willingness to defer to experts to explain why technology feels so out of control (Winner, 1983).

However, it is the term 'technological determinism' that is best known for conveying these ideas, even though it is a term that is both less appropriate and less consistently used. A real problem here, is that even amongst prominent and well-respected accounts of technological determinism, it is unclear exactly why they are so labelled, and what exactly is meant by determinism. A prominent example is the discussion provided by Merrit Roe Smith and Leo Marx, who distinguish between hard and soft technological determinism (Marx and Smith, 1994). In fact, the hard-soft distinction turns out to be polar cases of a spectrum of technological determinisms, with movement along the spectrum involving the degree of agency, or the power to effect change, attributed to technology. At the hard end, technology has certain intrinsic attributes that allow little scope for human autonomy or choice. At the other end of the spectrum, soft determinism simply emphasises the large scope for human interventions and choice. Indeed, for Smith and Marx at least 'the soft determinists locate [technology] in a far more various and complex social, economic, political and cultural matrix' (ibid., p. xii).

Why, however, should such accounts be considered to be deterministic at all? To the extent that both hard and soft versions accept that there is some scope for human choice, and merely contest the issue of how much, why would we want to use the term deterministic to describe either of them?[7] In response to this confusion of terminology, Bimber distinguishes nomological and normative forms of

---

[7] See also Lawson in Latsis, 2007.

technological determinism. The nomological is that which takes the 'determinism' in technological determinism most seriously. For Bimber, 'technological determinism can be seen as the view that, in the light of ... the state of technological development and laws of nature, there is only one possible future course for social change' (Bimber, 1996, p. 83). There is no scope for human desires or choices.

Alongside this form of determinism, Bimber distinguishes a version he terms 'normative' technological determinism. In the normative version, technology appears to us as autonomous because the norms by which it is advanced are 'removed from the political and ethical discourse and ... goals of efficiency or productivity become surrogates for value-based debate over methods, alternatives, means and ends' (ibid., p. 82). Here technological development is an essentially human enterprise in which people who create and use technology are driven by certain goals that rely unduly on norms of efficiency and productivity, thus excluding other criteria (ethical, moral) and producing a process that operates independently of the political processes and mechanisms usually thought to operate. The end point is one in which society adopts the technologist's standards of judgement. Thus there is a technological domain, which includes elements of society generally, acting as a constraint and a causal force on other aspects of society.

Searching for examples of such technological determinisms proves to be a revealing exercise. Notably, it is actually very difficult to find examples of the nomological technological determinism distinguished by Bimber. Most possible candidates (perhaps unsurprisingly) come from the economics domain. The most familiar of these is Marx's famous statement that 'the hand-mill gives you society with the feudal lord; the steam-mill society with the industrial capitalist' (Marx, 1956, 1955 [1900]). However, it is very difficult to attribute anything like a hard or nomological form of technological determinism once a wider reading of Marx is undertaken (see Chapter 11 or Rosenberg, 1976, Mackenzie, 1984a, Harvey, 2006).

Turning to Bimber's normative form of technological determinism, Habermas is singled out as a particularly good example (Habermas, 1970). Habermas bases his account on the distinction between work, which is success oriented – purposive action concerned with controlling the world – and interaction, which involves communication between subjects in pursuit of common understanding. Modernity is charac-terised by the colonisation of the system of objectifying (de-linguifying) behaviour of the former on the latter 'lifeworld'. Thus the problem, with which determinism grapples, is actually one of the inappropriate extension of one domain to another.

Although Bimber fails to mention him, a perhaps more obvious example of the normative form is provided by Heidegger. Heidegger, famously has a conception of technology that involves unavoidable negative change, ushering in a 'dystopian modernity'. For Heidegger, we are engaged in a transformation of the entire world (and ourselves) into mere raw materials or 'standing reserves', objects to be controlled (Heidegger, 1977, p. 183). Methodical planning comes to dominate, destroying integrity and encouraging a view of everything in terms of functionality rather than a respect for things for their own sakes. The central point is that technology itself is not neutral. The domination of technological processes leads to a situation in which everything is reduced to the status of a resource that has to be optimised in some way. Especially disturbing is the tendency for people to see themselves in the same way. Increasingly, sight is lost of what is being sacrificed in the move to utilise human and other resources for goals that become increasingly unclear.

Although many differences exist in Heidegger and Habermas's contributions, their central concern is strikingly similar, namely, the reduction of meaning and value in the domain of everyday living or the *lifeworld* that comes about through our engagements with technology. Underlying the accounts of Heidegger and Habermas is the idea that an instrumental attitude is adopted towards means and ends that results in various activities, including non-technological activities, being drained of meaning. Personal or emotional involvement is reduced to a minimum and the values of possession and control end up dominating social life. Our engagements with technology thus end up transforming us; the use of technology creates a new lifeworld, which isolates and impoverishes both the natural world and ourselves. Moreover, it is easy to see with these examples how the themes of evaluation and determinism are often connected.

Although there is far more to the philosophy of technology than can be discussed in such a short space, the above does serve to introduce the main ideas that this book is concerned with. However, as suggested above, it is important also to convey something of the rather peculiar status of the philosophy of technology as a discipline. More specifically, two overlapping questions require attention. The first is, why has the philosophy of technology only recently become formalised into a 'respectable' discipline? And, secondly, why have the main questions and themes of the philosophy of technology declined in importance as the discipline has become established? These questions are addressed in turn.

The Late Appearance of the Philosophy of Technology

Until recently, philosophers have been relatively uninterested in anything we may now refer to as technology (Dusek, 2006). The Research in Philosophy of Technology Society, the prime organisation for the philosophical study of technology, dates back only to 1975. And, within analytical philosophy at least, the study of technology has been viewed as at best a rather uninteresting 'specialisation'.

Several factors explain this state of affairs. One problem is the difficulty involved in specifying exactly what the term 'technology' refers to. Even a very casual review of the technology literature reveals the fact that different writers use the term 'technology' in very different ways. Technology is frequently portrayed as knowledge, as artefacts, as ways of doing things, as any means to an end, as a form of study and even as a form of social institution. Sometimes languages or symbolic devices such as calculus are treated as a technology. As noted above, in economics, technology is understood as the relationship between inputs and outputs or even as capital. How are we to adjudicate between such uses? Moreover, whilst there appears to be considerable agreement that certain things can be identified as examples of technology (such as computers, washing machines, aircraft, cameras, etc.,) and others that cannot (small children, flowers, paintings, jewellery, food, toys, etc.,) there is little agreement about which features of each grouping are responsible for the contrast.

If it is difficult to agree upon what the term technology refers to, how is it possible to discuss the broad features of societies that incorporate relatively more or less technology? Indeed, are such questions even meaningful? At least part of the disinterest in the philosophy of technology, or its questions, comes from a wider consensus that in the absence of a clear definition of technology, such questions are not really meaningful.

Even where a particular definition of technology has been applied, there are reasons why the philosophy of technology has not been taken very seriously. For example, one prominent definition has been of technology as 'applied science'. But if this is all technology is then, it would seem, the interesting philosophical issues lie elsewhere, such as in the philosophy of science, especially in the epistemology of science. Moreover, for some there is a belief that science, and so by implication technology, is in some sense 'neutral'. Thus, if technology has implications for society generally, given its unproblematic source in science, such implications are relatively uninteresting (philosophically) or can be dealt with independently as a matter of ethics (see for example Scharff and Dusek, 2003). As such, it is possible to see how the study of technology has tended to fall 'through the cracks' between science and ethics.[8]

Another likely reason for the philosophy of technology's particular status within the academy is the fact that it is, as a discipline, somewhat *ex post*; although a growing literature now exists setting out broad themes focused upon, and providing lists of examples of writers who can be thought of as contributing to a philosophy of technology, such demarcations would not have occurred to those authors in question (for examples of recent excellent anthologies see Scharff and Dusek, 2003, Dusek, 2006, Kaplan, 2009, Meijers, 2009). This might not in itself be a problem as long as such retrospective boundary setting captures something important, and I believe it does, but it has not helped the credibility of the (ostensibly) new discipline.

A further problem for the philosophy of technology has been the rather daunting range of authors and of topics that are now regularly listed as contributing to the discipline. This list, ranging from Socrates, via Heidegger, to recent actor network theory, not only requires a good knowledge of a wide range of thinkers, but also a familiarity with complex and diverse sets of ideas and themes.[9]

Despite these problems, however, the philosophy of technology is, as noted, currently receiving significant attention. However, as also

---

[8] One might also add that the truly worrying effects of technology, such as nuclear war, environmental degradation, biological manipulation, etc., have only emerged relatively recently.

[9] Indeed the complexity of the subject and the vast range of skills required have been suggested by some notable technology writers to account for the reason why technology is 'allowed to run out of control' (Ellul, 1980).

noted, this attention and academic presence has coincided with a fall in interest in the 'classic' questions identified at the outset as traditionally constituting the core of the new discipline. This claim requires some elaboration.

Criticisms of the Philosophy of Technology

By far the most sustained and prominent criticisms have come from constructivist positions, and are aimed at preconceptions of the philosophy of technology's main contributors that are perceived to be deterministic.[10] Although constructivism is itself far from being a homogeneous position, various points of agreement exist, at least with respect to the charge of technological determinism, which turn out to be important for the account that follows.[11]

First there is an explicit rejection of the idea that technical change can be seen as on some fixed or monotonic 'trajectory'. Technological change is, rather, thought to be genuinely contingent and not reducible to some inner technological 'logic'. Secondly, the view of the relationship between science and technology (in which science is seen as the independent, rational, non-political source of technological ideas) is questioned, as is the idea that technological change leads to (determines) social change and not vice versa. Instead, a rich source of case-study material is drawn upon to demonstrate the contingent nature of technical change and on how technology is 'shaped', especially by different social groups in the process of settling a range of technological/social controversies and disagreements (MacKenzie and Wajcman, 1985a). Crucially, at any point, there are many different trajectories or

[10] Prominent examples that do not quite fit into the constructivist mould tend to argue (in different ways) that the philosophy of technology focuses too much upon the effects of technology without considering very carefully what technology is (see for example Pitt, 2000, Mitcham, 1994, Kroes and Meijers, 2000).

[11] The main roots of social constructivist accounts of technology appear to lie in the sociology of scientific knowledge (for example, Bloor, 1976, Shapin, 1982). The term 'social constructivism' is most often used in a narrow sense to refer to the social construction of technology (SCOT) approach outlined by Pinch and Bijker (1987) or Bijker (1995) and related approaches (Woolgar, 1991). However, it is possible to include here the 'social shaping' approaches (MacKenzie and Wajcman, 1985b, Wajcman, 1991) and the actor-network approach of Latour (1987) and Callon (1987). These approaches are roughly in agreement (to varying degrees) over the following points.

routes that technology could take (it is underdetermined) and always depends to some degree upon human (especially collective) action.

Perhaps the greatest strength of these contributions is the plethora of demonstrations that technology is indeed irreducibly social. But there is much to say about the manner in which the social aspect is introduced, especially as regards the role played by the ideas of symmetries.[12] A central idea here is taken from the sociology of scientific knowledge literature (see, for example, Bloor, 1976): that it is best to remain agnostic about the truth, falsity, rationality, etc., of competing claims in settling scientific controversies. Translated to the technological realm this means that the researcher should remain impartial with regards to the actual properties of the technology and their relative functionality or efficiency determining which technologies become 'settled upon' (Pinch and Bijker, 1987).[13]

Given the de-emphasis of the nature of technology itself, in having much bearing upon its own acceptance or dominance, existing technology is understood or analysed in terms of the *stabilisation* of different controversies and disputes. Once stabilisation is achieved, controversy is removed and the properties of this stabilisation (how consensus is achieved) determines how that technology functions. The focus, as with the sociology of scientific knowledge literature, is upon how 'closure' is achieved. Crucial to this conception of closure, is the idea that technology is not interpreted or understood in any fixed way. These different interpretations of technology are not only concerned with its social characteristics or relative functionality, but with its technical content – with how it works. Thus, 'facts' about technology are simply the (different) interpretations of different social groups

---

[12] Various accounts put this aspect as central stage (Lynch, 1992, Collins, 1985). Some accounts even present the field as a series of extensions of the symmetry principle (Woolgar, 1988).

[13] The researcher must, in other words, treat as possibly true or false all claims made about the nature of technology – such claims must be treated symmetrically (explaining them by reference to similar factors), since there is no independent way of evaluating the knowledge claims of scientists, technologists, et al. As in the sociology of scientific knowledge literature, two ideas underlie these arguments. First, the 'real world' plays no role in settling controversies (in settling the form that technology takes) and, second, that the researcher has no independent access to the world, so there is no way of evaluating competing claims. Thus claims about the relative efficiency or successfulness of different technologies or technical progress (or how some technology comes into being) are to be avoided (Staudenmaier, 1995, Pels, 1996).

(Bijker, 1995). The term 'closure' is therefore used to refer to a rhetorical process of settling disputes via negotiation and social action. Technology is thus socially shaped and socially constructed.

The problem with these responses to technological determinism is that they result in a situation where it is difficult, if not impossible, to distinguish technology from any other social phenomenon (Winner, 1991, Lawson, 2007). Whilst there may be some questions or analyses within which this is itself not a problem, it certainly makes it impossible to address any of the 'technology questions' noted above that form the core of the traditional philosophy of technology. For example, it makes it difficult to explain the special role that technology takes in prodding or provoking all manner of social changes.

Constructivists are clearly correct to argue, against nomological determinist accounts, that contingency matters. However, little is said about the normative form noted by Bimber and others. Unseating the privileged position of science in technology's development, whilst also surely correct, actually distracts from the very factors to which Habermas, Heidegger and others are drawing attention – that the use of technology tends to bring with it values of possession and control that can dominate social life and drain it of meaning. These are questions about which the constructivist has little to say.

Put bluntly, reasonable criticisms of aspects of the determinist position have encouraged a situation where technology cannot be distinguished from any other social phenomenon, and so the classic questions of the philosophy of technology become impossible or unnecessary to address. In contrast, this book is concerned with just these questions. However, I shall argue that such concerns are quite compatible with the irreducibly social character and essential contingency insisted upon by constructivist accounts. Moreover, I shall argue that these questions are most easily addressed by establishing a conception of technology that systematically draws upon and develops recent accounts of social ontology.

Outline of the Book

To take stock, I believe there is much of value in the older, or classic questions of the philosophy of technology. And whilst there is also a good deal of value in the more recent, predominantly constructivist, criticisms of this literature, constructivist critics end up painting themselves into a corner in which it is impossible to address, or even pose, the main questions with which the philosophy of technology has been most concerned. In short, it is impossible to portray technology as being different from any other social phenomena.

This being the case, there is a need to find a way of formulating these older questions fruitfully, which accommodates the more interesting constructivist criticisms of technological determinism. In order to do this, as noted above, I wish to draw upon relatively recent developments in social theory, specifically social ontology, which have thus far not found their way into the technology literature in any sustained manner. In short, my aim is to make an ontological contribution to the philosophy of technology from a particular social ontological perspective.

In order to facilitate this, some introductory and context-setting remarks are required. This is the task of the first part of the book. More specifically, the following chapter provides an account of the nominal features of technology that require accommodation. This is pursued by asking when and how the term 'technology' acquired its current significance. Chapter 3 makes the case for pursuing an ontological approach and also sets out the ontological strategy I adopt, along with a summary and development of recent ontological contributions that I draw upon in later chapters.

Chapter 4 begins to construct an ontological account of technology by first considering the relationship between science and technology. Although critical of classic conceptions of technology as applied science, this chapter maintains that understanding the nature of science is still important for understanding the nature of technology. Chapter 5 focuses upon the irreducibly social nature of technology highlighted by constructivists. However, this 'social-ness', or sociality, of technology is formulated in more ontological terms, especially in terms of social positions. Chapter 6 attempts to distinguish technological artefacts from other kinds of artefacts, and Chapter 7 argues that such artefacts can also be understood to be extending human capabilities. Chapter 8 attempts to consolidate ideas from the previous four chapters on the ontology of technology by making comparisons to other recent accounts of technology, especially to instrumentalisation theory.

The final three chapters draw out implications of the account given for a range of very different questions. Chapter 9 focuses upon the relationship between technology and those with autism. Chapter 10 considers the dynamics of technological change, especially focusing upon the claim that modern societies are in some sense speeding up. The final chapter returns to ideas set out above, arguing that a rejection of technological determinism does not imply a form of technological neutrality; there is much that can be said about the general nature of technology without committing oneself to any form of technological determinism.

2 | From Obscurity to Keyword: The Emergence of 'Technology'

As noted in the previous chapter, any attempt to discuss the nature of technology is confronted by an immediate problem. Despite a widespread acceptance that the term 'technology' refers to something central to modern societies, there seems to be little appetite for the task of either formulating a precise meaning of the term or for prioritising one particular existing usage amongst others. In fact, there is often little agreement about what *kind* of thing the term technology refers to; it is routinely used to refer to material objects (some, but not all, of which have been transformed by human activities), practical or scientific knowledge, inventions, applied science and knowledge embodied in things (often material objects but not always), particular practices and even social institutions (Faulkner et al., 2010). How is it possible to proceed when such widely different usages of the term are commonplace?

In order to formulate, and in part defend, the definition of technology that underlies the discussion of later chapters, I shall focus upon the circumstances in which the term emerged. More specifically, in this chapter I shall be concerned with accounts by historians of technology that relate to a particular historical episode in which the term 'technology' not only underwent dramatic changes in its usage but also became a popular and significant term both in social theory and everyday discourse. In particular, older, more etymologically faithful, understandings of the term began to be transformed around the middle of the nineteen century, giving rise to various overlapping uses which nevertheless started to stabilise in the middle of the twentieth century. In this chapter, I trace out some of these transformations, focusing in particular on the events that prompted changes in the term's usage. I also highlight some continuity throughout these changing usages of the term, arguing that in fact, such usages can be understood as extending existing meanings rather than replacing them.

On a terminological note, most of the accounts to which I refer here have been motivated (sometimes explicitly) with a concern for what

Raymond Williams terms 'keywords' (Williams, 1976). The main idea
here is that when prominent new words come into being, the contests
that result in these changed meanings, and the emerging actors who
drive these changes, tell us much about the historical forces at play in
a particular period. In short, keywords act as windows of epistemic
access onto different historical epochs. Although such windows are
also significant for the account I wish to develop, the emphasis I adopt,
as well the main motivation for focusing upon particular developments
are, of course, rather different. My main interest is in the emergence of
a particular keyword, technology, as a basis for developing an ontological account that can accommodate the main features that those using
the term were concerned with, so hopefully making my account
persuasive. Put another way, the term 'technology' emerged as a keyword in social theorising at a particular historical juncture and in
reaction to a particular set of issues. Any worthwhile ontological
account of technology should be able to accommodate and explain
those features.[1]

Historical Meanings of 'Technology'

As noted, there is general agreement that 'technology' emerged as an
important term in public discourse around the beginning of the twentieth century. Before this point in time, the term was used pretty much
in line with its etymological roots. These roots lie in the combination of
the two Greek terms, *tekhne* (skill or art) and *logos* (discourse or
study). By the early nineteenth century English, French and German
dictionaries and encyclopaedias were largely in agreement in drawing
upon these roots to define technology as the 'description, principles or
teaching of the practical or useful arts'. Most notably, they agreed that
technology was a field of study, not an object of study. When used in

---

[1] One last preliminary problem is that, if use of the term 'technology' became
widespread only relatively recently, dating roughly from the beginning of the
twentieth century, how is it possible to talk of the philosophy of technology as
stretching back to ancient Greek societies? Pointing this out, however, simply
serves to underline the fact that the philosophy of technology has been formalised
retrospectively, in the sense that recent contributors have projected a concern
with what we now call technology back on to earlier societies and earlier
contributions. The question then becomes when did the term, which we are now
projecting back to previous epochs, come into being and why? Or, when, and
why, did the term become a significant component of analysis in social theory?

this way, 'technology' remained a very obscure, specialised term that was in practice used very rarely in any language; although there was some flexibility in its meanings, none of these meanings seemed of particular interest (Schatzberg, 2006).[2]

By the middle of the nineteenth century this very specialised use of the term gave way to something quite different. Initiation of this change is often attributed to Jacob Bigelow. Writing in 1829, Bigelow, a medical doctor and Harvard professor, used the term in the title of his lectures on applied science (*Elements of Technology*). In his preface, he made such grand claims for the extended applicability of this unusual word that for well over a century historians mistakenly credited him with having anticipated the latter-day meaning of technology (L. Marx, 2010). On closer reading, however, Bigelow's usage of the term was not so very far from earlier usages (to name a particular field of study – in particular the 'mechanic arts').[3] And it is probably more appropriate to conclude that Bigelow's contributions remain important more for changing the status or profile of the term, even if the term's meaning was left poorly defined.

To understand how the modern use of the term emerged, we have to follow several different developments that came together in the early years of the twentieth century, predominantly in the United States. The first set of developments revolve around the work of Thorstein Veblen, the second concerns the changing role and status of science and engineering, the third relates to changing conceptions of progress, and the final development is the greater structural integration of new inventions and innovations leading to the prominence of what have come to be termed sociotechnological systems. Although I shall initially discuss each theme separately, as I make clear, they all overlap and reinforce each other.

---

[2] One exception to this may be the account of technology provided by Karl Marx, especially in Book 1 of *Capital*. Whilst it is clear that Marx developed a fairly sophisticated conception of technology, predating many of the developments in the term that were to follow in the early twentieth century, he failed to use the term very consistently. Moreover, his usage of the term does not seem to have directly affected its mass adoption in the early part of the twentieth century. For these reasons, and given that Marx's contribution forms the focus of a later chapter, Marx's contribution is not examined here.

[3] Moreover, Bigelow, having resurrected the term, promptly dropped it in the second edition of his book (retitled *The Useful Arts*). Though Bigelow, through his influence on the Massachusetts Institute of Technology, did have some impact on the spread on the term in later years.

Veblen and 'Technik'

Veblen's significance in the emergence of the term 'technology' comes in large part from the role he played in translating ideas taking place in Germany from the mid to end of the nineteenth century (Misa, 2009). Like the United States, Germany was experiencing rapid industrialisation in this period. In this context, a set of debates emerged in relation to the concept of 'technik', a concept that was rarely ever compared to the older use of the term as a form of study (Schatzberg, 2006, p. 494). In English, 'technik' is most often translated as the term technique (of a painter or musician). But in the German discourse of the time, technik had a complex dual meaning, referring both to the material components of industry as well as to the rules, procedures and skills used to achieve some particular end (Mitcham, 1979). This double meaning made it a very useful term. For example, as is discussed more fully in the following section, technik's double meaning played an important role for engineers who were keen to upgrade their professional status. In particular, the dominant engineering organisation in Germany, the Verein Deutscher Ingenieure, embraced technik as its central concept in its fight for improved social status within the German social hierarchy (Gispen, 1989).[4] This desire for status was almost always couched in terms of the importance of technik (Schatzberg, 2006). This in turn, generated a full blown debate about the relationship between Technik and Kultur, which greatly influenced the development of German social theory, especially for some of the prominent members of the German historical school such as Gustav Schmöller, Max Weber and Werner Sombart.

Veblen's work was by far the most important route by which this debate found its way to the United States. The term 'technology' was first linked to the term 'technik' when Veblen undertook reviews of the work of Schmöller and Sombart, and the term became a major conceptual component of Veblen's thinking soon afterwards. Central to Veblen's contribution is his analysis of the relationship between business enterprise and modern industry. Within this account, the term 'technology' is typically used to refer to a set of socially beneficial tendencies that stands in opposition to a set of parasitic or harmful ones. On the 'good' side, there is workmanship, industry,

---

[4] This concern with status was wide ranging, including for example protracted discussions of whether engineers should be required to learn Latin.

the development of industrial machinery (the 'machine process') and practical knowledge. On the other side is predation (including hostile takeovers), business enterprise, absentee ownership and other 'pecuniary' institutions. The nature of this distinction is taken up in more detail in Chapter 6. For now, two points are important to make for the reformulation of the term 'technology' in Veblen's hands.

First, Veblen developed a wide ranging (if not always well understood) account of the relationship between science and the kind of knowledge that featured in earlier uses of the term 'technik'. Scientific and 'technical' knowledge can be traced to two distinct instincts, idle curiosity and workmanship, which develop alongside each other. Idle curiosity becomes the basis of modern science, whilst workmanship drives the shift from handicraft to the machine process in Veblen's account. For Veblen, these processes mutually facilitated each other. In particular, the machine process, which highlighted a shift from the dominance of craft to machine-based activities, instigated different conceptions of science itself. Specifically, Veblen argued that very basic conceptions of how explanation is viewed or understood changed dramatically (primarily moving away from vaguely defined, often theological explanations of phenomena to hard-nosed conceptions of cause and effect), so significantly shaping modern science.

The second point is that the term 'technology' became a central component of Veblen's critique of capitalism. Veblen argued that technological knowledge (which for him includes language, the use of fire, tools for cutting, etc.,) had always been integral to human communities and constituted their 'immaterial equipment' of production. Such knowledge was cumulative and collective, in that it belonged to nobody in particular and was transmitted through community use. General increases in well-being or development came about when both the material and immaterial developed together. But, for Veblen when the material component was in short supply, certain individuals could in effect monopolise the community's collective technological knowledge by controlling the material means of utilising this knowledge, therefore creating the basis for pecuniary domination (Veblen, 1908). What Karl Marx terms the means of production represented, for Veblen, a theft of the community's collective technological knowledge.

Veblen's use of the term technology is perhaps most developed in the Engineers and the Price System (1921). Here Veblen introduces his famous idea that production engineers should be in the vanguard of the overthrow of the vested interests. This soviet of technicians would be made up from skilled experts drawing on the collective stock of technological knowledge. Although Veblen is often credited with popularising a conception of technology centred on engineers, a closer reading reveals that this is not Veblen's intention. For example, Veblen certainly did not identify his soviet of technicians with the existing engineering profession, which (in the process of professionalising) Veblen clearly identifies as subservient to the 'vested interests'. But as so often with Veblen, distinguishing his serious points from his satire is not always straightforward.

Whatever the precise understanding of Veblen's strategy for overthrowing the vested interests, it should be clear that his use of the term 'technology' had now drifted far from the nineteenth century meaning of the term. Although not always consistent, his use of the term included technical knowledge and the practice and material basis for the industrial arts and was firmly independent of science. A crucial step in Veblen's argument is that in elevating the status of engineers in an account that used useful and productive (rather than wasteful or pecuniary) as criteria for 'the progressive', engineers had to produce something. In this context, technology emerged as a term that not only denoted what it was that engineers produced, but which emphasised the material, typically machine-like properties of that product.

In a matter of years, Veblen's arguments were influential in the technocracy movement, especially in its focus upon the major debates about technological unemployment (in particular, the displacement of labour by machines). This movement was consciously not Luddite, in that advocates did not call for an end to mechanisation, but it did lobby to spread the costs of this unemployment as equally as possible. Importantly, however, most advocates of the movement were critical of industrial capital and drew on Veblen's work to call for a restructuring of the price system and government by engineers. As Veblen had correctly guessed, the established engineering community wanted nothing to do with such ideas. Explaining quite why this was the case and how the status of engineering was changing is another crucial part of the story of technology's shifting meaning.

The Changing Status of Science and Engineering

Various historical accounts, concerned with tracing out the emergence of the term 'technology', highlight the importance of changes in science and engineering at the beginning of the nineteenth century that to some extent mirror the earlier German experience. An important example is provided by Kline (1995). Whilst covering a wide range of views on the nature of the relationship between science and inno-vation, Kline argues forcefully that these differences were subsumed under 'the flexible rubric of applied science in the process of drawing boundaries and promoting complementary self-interest' (Kline, 1995, p. 198). In particular, scientists tended to emphasise the dependency of practical innovations upon pure research to argue for the financial support of their work. At the same time, engineers called themselves applied scientists to raise their occupational status above that of artisans to that of learned professionals. However, both groups (scientists and engineers) required a new, legitimating word for the output of applied science. The word that seemed most appropriate was 'technology'.[5]

Ruth Oldenziel traces a similar story of the vying for status and redefinitions of disciplinary boundaries. However, her focus is upon the ways in which these groups used the new term 'technology' to systematically exclude the contributions to inventiveness and industrial development of women and non-whites. Oldenziel suggests that the relationship between technology and masculinity was essentially constructed in the early part of the twentieth century when the male-dominated engineering profession appropriated the term. Oldenziel, in line with many others, suggests that the term replaced reference to the study of the useful arts, but she links this replacement to the construc-tion of gender and class relations. Before the twentieth century, a reference to the useful arts would have included agriculture, needle-work, baking, etc. But as the century went on, different historical actors began to classify and lay claims to a range of useful objects, arts and knowledge, privileging some and discarding others. Machines were put centre stage 'as the measures of men and markers of modern manliness' (Oldenziel, 1999, p. 19).

---

[5] These transformations are particularly well highlighted by Ellen Swallow Richards, who observed in 1911 that MIT's stated purpose of applying science to the useful arts became respectable in academia because the school focused on technological, rather than technical, training. 'Technology is the incorporation of higher scientific knowledge into the arts, a process that is now taking place to such an extent that one may almost say the "science of yesterday is the technology of today"' (Swallow Richards, 1911, p. 126), quoted in Kline (1995, p. 217).

Oldenziel draws on a wide variety of sources to make her case. For example, she is concerned with showing how in a series of different contexts, from world's fairs to the restructuring of education establishments, the term 'technology' emerged as a label of self-identification for ambitious engineers and applied scientists. Elsewhere she draws upon the work of Lewis Morgan, which focused on the extent to which the rate of innovation might be a prime mover of societal evolution, detailing how 'machines came to function as measuring devices by which western cultures gauged themselves with increasing confidence and assessed other cultures with condescension' (ibid., 42). Quilts and corsets, she argues, all important objects of women's inventive activity throughout the nineteenth century, were increasingly filtered out of contemporary minds as markers of true technology.

Through the prism Oldenziel constructs, Veblen's contributions, however intended, further cement such developments. In demarcating between the useful and the pecuniary, Veblen reinforced a conception of 'male-machines' and thus a particular set of gender relationships. His discussion of women of the leisure class is significant here, women being viewed by Veblen as useless and idle, serving only as tokens of the status of men.

As noted, such arguments played an important role in the formation of the technocracy movement and the debates about technological unemployment, where technology first became a household term (Kline, 1995, p. 197). As also noted, the established engineering movements wanted nothing to do with the technocracy movement, and deflected such associations by focusing upon the benefits of industrialisation. This response saw a range of developments including a dramatic change in emphasis upon the importance of consumption rather than production (e.g. as witnessed in the restructuring of the Chicago World's Fair in 1933, see Oldenziel, 1999). Throughout this period the term 'technology' rose to prominence both as a central component of the technocracy movement, but also in the reaction against it. Importantly, with both movements came the first significant attempts to link the new term technology to that of progress.

Technology and 'Progress'

The idea that this link to progress is the crucial step in understanding the transformation in meaning of the term technology at this time, is a position held by many historians of technology, but the case is perhaps most forcefully made in the work of Leo Marx (see especially L. Marx, 2010, L. Marx and Mazlish, 1996, L. Marx and Smith, 1994). Marx's main argument is that a number of societal changes created a semantic or conceptual void. Whilst Carlysle ably summed up the first half of the nineteenth century by referring to the 'age of machinery', such a label did not quite fit the latter half of the nineteenth century and onwards. The important question is, then, why?

Marx's answer is that the prominence of the term 'technology' resulted from a combination of the awareness of (or indeed a preoccupation with) progress, with the acknowledgement of the power of innovations to transform both the world and how that world is interpreted. Here again, Veblenian themes (this time of changing habits of thought) are important, but Marx raises other interesting themes as well. From about the 1700s onwards the term 'progress' was used to refer to incremental but largely discrete advances in the useful or material arts, such as developments in new scientific instruments. By the time of the French and American Revolutions, history itself was conceived of as a record of steady cumulative and continuous expansion in the power of humans over nature, along with an improvement in the conditions of human beings. But by the mid-nineteenth century there was a subtle change in this 'master narrative'. For Condorcet, Paine, Franklin and Jefferson, human progress was not simply equated with the advance of the mechanic arts. They were committed republicans and political revolutionists. Although they celebrated mechanical innovation, they celebrated it only to the extent that it helped further different aims – 'the true and only reliable measure of progress, as they saw it, was humanity's liberation from aristocratic, ecclesiastical, and monarchic oppression, and the institution of more just, peaceful societies based on the consent of the governed' (L. Marx, 2010, p. 565). For these writers, advances in science and the mechanic arts are valuable chiefly as a *means* of arriving at social and political ends.

By the mid-nineteenth century, Marx argues, such advances did more than facilitate progress, they embodied it or simply *were* progress. As John Stuart Mill acutely observed, 'the mere sight of a potent machine like the steam locomotive in the landscape wordlessly inculcated the notion that the present was an improvement on the past, and that the future promised to be wondrous'.[6] The term 'technology' came to be associated with this embodiment of progress in the results of engineering and applied science, especially as such embodiment came to be an explicit dimension of some of the social theories emerging at that time.

The technocracy encouraged by Veblen's work is one example. But another is the rather deterministic conception of material progress offered by Charles Beard. Beard combined an enthusiasm for reform with a profound faith in the progress of civilisation. The material developments of the time were the main source of this positive outlook. But Beard did more than view such inventions as the steam engine, as visible signs of progress. He also explicitly linked the term 'technology' to the idea of progress in a way that made technology itself the motive force of history (Schatzberg, 2006). Although drawing upon Veblen's ideas, his conception of technology had none of Veblen's subtlety. However, Beard's account proved enduring, especially in suggesting that technology was a relatively independent force which was to be welcomed as an agent of beneficial change.

This blurring of the distinction between mechanical means and political ends, which takes such an explicit form in Beard's contributions, served also to generate an ideological backlash. For a small but noisy minority of dissidents (often artists and intellectuals), the elevation of the status of material progress was seen as symptomatic of a moral negligence and political regression. Indeed, it seems possible to draw a more or less direct line of influence from the intellectual dissidents of the mid-nineteenth century to the widespread political demonstrations of the 1960s, to de-growth contributors of the present day, which were and are primarily a revolt against an increasingly progress-obsessed, and often technocratic, society (L. Marx, 2010, p. 567). The main point that each group agreed upon was the idea that such forces could not adequately be captured simply by referring to machines or material devices. Rather a new term was required that could capture the coming together of practical knowledge, innovation and material devices in such a way as to drive profound changes to social life. The term that each group adopted was of course 'technology'.

---

[6] John Stuart Mill (1840) quoted in L. Marx (2010).

Complex Sociotechnical Systems

To this blurring of material innovation and political ends, might be added a change in the nature of material innovation and in the way innovations are 'put to work'. In particular, the period in which the term technology emerges and becomes linked to ideas of progress also sees an increasing structural complexity in the nature of the material innovations in question. For example, during the early phase of industrialisation, innovations in the mechanic arts tended to involve more or less free-standing or self-contained mechanical devices, such as the power loom, the spinning jenny, the dynamo or other machines.

By the middle of the nineteenth century, the major innovations tended to involve wider systems of interlocking mechanical infrastructure, or sociotechnological systems (Bijker et al., 1987, Hughes, 1983). Here new systems emerged where the central defining artefact was actually a relatively small, although crucial, part of the whole. An obvious early example is the rail system. Although the steam locomotive was an essential component, the working of the system involved all kinds of supporting artefacts, such as stations, rolling stock, bridges, tunnels, signalling systems, and huge networks of tracks. It also required huge amounts of capital investment and a very organised management structure, specialised knowledge and skills, a specially trained workforce, regulations concerning gauge widths, etc.

Within such newly emerging systems there was essentially a blurring of the components, in particular between the artefactual equipment and everything else. Moreover, unlike those innovations associated with the eighteenth century Industrial Revolution, which had often been introduced by relatively isolated enthusiasts and practical mechanics with relatively little formal scientific training, many of those who developed and refined the workings of these large systems had been educated in establishments in which the vying for status, noted above, was a central concern (Bijker et al., 1987). This latter phase, often referred to as the second Industrial Revolution, or simply the technological revolution (Smil, 2005), stemmed more directly from advances in science that became centrepieces of large complex systems in which people and material artefacts combined. Thus, making technological artefacts 'work' involved enrolling in them in often-extended networks of interdependencies. The term 'technology', then, was not only required to accommodate a focus upon the outputs of material innovation, but the resulting networks in which such material innovations were to operate.

Defining Technology

There are many important features of the above account that are returned to in later chapters. Here, the main concern is with the common ground in these accounts, and in particular the idea that this period saw a transformation in the meaning of the term 'technology' away from the study of an object to that of an object of study. Each of the accounts referred to above suggests that the earlier, more etymologically faithful understanding of technology as the study of the useful arts was rarely used and only of specialist interest. The term only really gains any general interest as its meaning became transformed, somewhere between the middle of the nineteenth century and the 1930s and 1940s. Veblen's work, in translating the German ideas surrounding the term 'technik', and in driving a wedge between modern capitalism and the progressiveness of technology, had a crucial part to play in popularising the term. Also of importance to the transformation of the term's meaning were the contested, but often complementary attempts of scientists and engineers to carve up institutional boundaries, which might allow for the intellectual authority of science but also the growing professionalisation of engineering. Related to this were the often insidious attempts by both groups to exclude unwanted 'others' such as women and non-whites. In the complex interactions of these developments there was a growing need for a new term that had to perform various tasks. Centrally, the new term had to refer to not only developments and applications of knowledge, but to the material results of such developments and applications and their practical realisation.

Technology emerges as a keyword in order to accommodate all these developments. However, although there is significant agreement amongst historians of technology that this rise in the term's popularity coincides with the need to talk about an object of study rather than the study of an object, it can also be argued that the complex shifting of meanings to be found suggests not simply that one meaning replaces another. Rather, the picture that emerges is one in which meanings are extended, rather than replaced, to such an extent that we end up with a definition of technology as both referring to a singular object but at the same time as accommodating other slightly different meanings.

More specifically, I want to suggest that technology can be understood in a way that preserves some of its etymological roots. As noted
above, technology was initially understood as a study, in this case of the
'description, principles or teaching of the practical or useful arts'.
However, it is also generally acceptable to talk of the results of different
forms of study using the same term. Thus, just as forms of study, such
as ontology, sociology, history, etc., generate results that are labelled
an ontology, a sociology, a history, etc., it is also possible to talk of the
results of the study of the useful arts as a technology.

There is though a difference between the results of technology and of
these other forms of study, which the accounts above highlight as being
of central importance in explaining the term's dramatic surge in use.
In the case of society, history and ontology, the objects of study exist
prior to being studied, and so are independent of the latter. In the case
of technology, however, the goal is ultimately to bring a non-existent
object or feature into being. Of course, it is possible to restrict the idea
of *a* technology to a set of ideational plans and/or schemes (for transforming or organising/combining objects) that result from technology
qua study. But mostly the process of study involves not merely ideas but
also the creation of novel objects themselves and assessing their properties via trials and experiments with them. As a result, it is reasonable to
allow the term technology to further extend to cover the material
results of technology *qua* study. Strictly speaking, though, any material
object or system so produced remains *a* technology (rather than technology). To minimise confusion, I shall often refer to the knowledge
produced in such procedures as technological knowledge, and such
material products of study as technological objects, or more usually
technological artefacts. But use of the term 'technology' also should be
able to include all of these features, study, knowledge and artefact for
the simple reason that it is the study and knowledge embodied in the
resulting artefacts that give them their particular character.

In sum, then, technology is best understood as having a multicom-
ponent meaning. Its roots remain as a form of study, often presented as
a study of ways and means, of the practical or useful arts. It is, secondly,
the ideational results of this study, any such result being referred to as
a technology, or technological schema or plan. But thirdly, and most
importantly, the term also covers features, objects, or structures that
materially result from and through such study. These features will
frequently be identified as such through use of the adjective

'technological' (as in technological artefact). Use of the term technology by itself, however, is intended to refer to the realisation of study and knowledge in particular artefacts, and thus is intended to refer to the whole process in which technological artefacts come into being and the results of studying how causal factors and insights identified in science (and elsewhere) can be practically harnessed.[7]

Let me also be more specific about the form of such harnessing. In particular, I argue that technology is primarily concerned with harnessing the capacities of material objects in order to extend human capabilities. Each of these terms needs unpacking and I shall return to each in later chapters.

Whilst this current chapter has been concerned with how the term technology emerged and with adopting a definition of the term that is in keeping with its rise in popularity, from now on I am concerned with ontology. Specifically, Chapters 4, 5, 6 and 7 are concerned with an ontological elaboration of different aspects of the definition adopted here. However, the term 'ontology' has a rather mixed history within the social sciences. Indeed, within the study of technology, ontology is often viewed with a good deal of distrust. Given my intention to provide an ontology of technology that can be both plausible and useful, I must first present the case that ontology is a useful concern. Moreover, I need to spell out which particular approach to ontology I shall be adopting, as well as signalling some of the advantages of adopting such an approach. This is the aim of the next chapter.

---

[7] See Chapter 4 for an elaboration of the argument here.

3 | Ontology and Isolation

Ontology is usually defined very broadly as the study of being or as the study of the kinds of things that exist. But we all know something about ontology and use it on a daily basis. To use a simple example, a stick may be a useful tool to clean a carpet, but it would not be a very good tool for cleaning a window. We know this because we understand something about the natures, the being, of windows, carpets and sticks.

Ontology, has enjoyed something of a resurgence in recent years. However, it continues to elicit very different, often polarised, attitudes in social theory. Some view ontology as either irrelevant, a 'fifth-wheel' that generates no practical implications or concerns, or (if it does have implications) as an undesirable form of essentialism, most likely betraying some unreconstructed form of 'modernism'. For others, myself included, ontology is viewed as essential and non-optional in the sense that all social theorising presupposes some kind of ontology, and it is better to be explicit about this than to remain wedded to an implicit ontology that may well be fraught with important problems and inconsistencies. It is fair to say that amongst those currently concerned with the study of technology, it is the former view that tends to dominate. The main task of this chapter is to argue for a particular conception of ontology that can be useful for the study of technology whilst avoiding the usual criticisms levelled at it.

More specifically, this chapter first provides some of the context to these differing views concerning ontology's significance. It then lays out a broad conception of ontology that, it is hoped, might meet the concerns of even those who tend to be suspicious or critical of ontology. In so doing, this chapter also introduces and develops a series of ontological arguments, concepts and distinctions that will be drawn upon throughout the remaining chapters.

Contextualising Ontology

The term 'ontology' dates back to at least the late scholastic writers of the seventeenth century.[1] The traditional understanding of the term is the science or study of being, deriving from the Greek 'onto' (being) and 'logos' (study or science).[2] Defined in this way, of course, ontology might involve the study of just about anything, and those who dismiss ontology as irrelevant often make just this point. However, in practice, the following narrower ontological concerns have tended to dominate. First, ontology is restricted to the study of anything under the aspect of its being, or in other words, the study of what is involved in something's existence. For example, waves, particles, rules, magnetic fields and debt all exist, but do so in different ways. Thus ontology is often portrayed as concerned with 'modes of being'. In light of this, ontology has tended to be distinguished from other domains of philosophical enquiry such as epistemology (the study of knowledge) and methodology (the study of method).

Secondly, ontology tends to focus on the study of those entities or phenomena that are regarded as the most basic or significant in some sense. Although such an orientation is sometimes viewed negatively, as the first step towards some form of essentialism,[3] such a focus actually tends to involve little more than a particular (and perhaps unsurprising) limiting of scope to those features of being that might be expected, in some historical context, to be of most interest.[4]

At least some of the negative attitude towards ontology arises because of its traditional association with metaphysics. Although, like ontology, metaphysics is not always well defined, a view persists that it is inaccessible to those without the 'correct philosophical training' and, in any case, quite irrelevant to the concerns of anyone other than

---

[1] 'Ontology', or rather 'ontologia', appears to have been coined in 1613 by two philosophers writing independently of each other: Jacob Lorhard in his *Theatrum Philosophicum* and Rudolf Göckel in his *Lexicon Philosophicum*. Its first occurrence in English seems to be in Bailey's Dictionary of 1721, where ontology is defined as 'an account of being in the abstract' (T. Lawson, 2015).

[2] At this level of generality, the word 'being' can itself be understood in two senses: first it refers to those entities or things that exist and, secondly, it refers to what it is to exist, i.e. to what (if anything) all things have in common.

[3] Where essences are usually understood rather dismissively as unchanging and deterministic.

[4] See, for example, Sayer (1997).

specialist philosophers.[5] Although I want to argue that ontology (and indeed metaphysics) should not be viewed this way, there is much in the term's history that does encourage this attitude. Let me elaborate.

Perhaps the first major contribution to ontology came with the publication of *Philosophia Primasive Ontologia* by the German philosopher Christian Wolff. Wolffian ontology was essentially an application of the method of deduction to philosophical problems generated by the Aristotelian distinction between substance and accident. Beginning with indubitable first principles such as the principle of noncontradiction, Wolff hoped to deduce the contents of the world without engaging in the messy task of empirical research. Though it persists today in some scholastic manuals, this tradition quickly lost its philosophical respectability during the eighteenth century principally as a result of the critical contributions of Immanuel Kant. In his *Critique of Pure Reason* (1781), Kant demonstrated that a priori methods could be used to deduce both the thesis and the antithesis of Wolff's arguments from the same premises, thus dealing a fatal blow to the earliest modern tradition of ontology. After Kant, although metaphysics survived in some form, the study of being *qua* being was treated with increased caution.

From the mid-seventeenth century onwards, the work of British empiricists reoriented philosophical investigations towards an emphasis upon human perception. These developments culminated in the twentieth century emergence of logical positivism and logical empiricism. Following the work of Hume and Berkeley, metaphysics came to be associated with unfalsifiable speculation. As such, the terms 'ontology' and 'metaphysics' became interchangeable and used pejoratively to undermine unpopular philosophical positions and attack opponents. As a result, the ontological dimensions of philosophical and scientific thought were rarely discussed.

---

[5] The term 'meta' in Greek means over or after and the term 'physis' translates as 'nature'. The term 'metaphysics' gained currency from Aristotle's *The Metaphysics* (*ta meta ta phusika*) which was placed immediately after the book called *Physics*. The term seems to have gained wide appeal as denoting the *purpose* of metaphysics, which is (or includes) reaching above or beyond nature (*physis*) to uncover its most basic components of fundamental features. It is this status as in some sense 'over and above' science which seems to lie at root of much of the deep suspicion that scientists have for metaphysics, often using the term to convey the idea of idle or detached speculation. See for example the essays on ontology in Burkhardt and Smith (1991).

The middle of the twentieth century saw a reversal of this trend and the reintroduction of ontology into mainstream philosophy. The driving force here was Willard Van Ormen Quine, whose collection of essays *From a Logical Point of View* (1953) drew attention to the crucial role of ontological commitment in the construction of both scientific and philosophical theories. It remains unclear, however, how much Quine's project really breaks away from the tradition of logical empiricism. On the one hand, Quine argues that ontology requires consideration of the presuppositions of scientists; ontology must be concerned with what scientists are committed to. But on the other hand, Quine, in accepting the reliability of (at least natural) scientific accounts, really does believe that he arrives at knowledge of the external world. Certainly, philosophers following in Quine's steps, have tended to assume that natural science is the main candidate for our best epistemic practice and hence that we should look to the natural sciences for our ontological commitments.

In the work of later ontologically-oriented philosophers of science such as Harré and Madden (1975), Bhaskar (1978), Hacking (1983), Fine (1986), Cartwright (1999) and Ellis (2001) there has been more attention given to the nature of science as an ongoing, productive activity. These authors, moreover, continued the work of reintroducing ontology into philosophical discourse by emphasising that scientific activity presupposes particular ontological conceptions and that these are (in some sense) implied by our best scientific descriptions of the world.

Ontology for Social Science

Such developments have found their way into social science through a variety of means. One is through the use of formal philosophical methods to the study of different sciences, amongst them social sciences. However, such philosophy of social science has, following part of Quine's concerns, tended to focus upon issues of truth, verisimilitude and rationality, rarely making commitments to a specific ontological outlook. Contributions have for the most part been made by those considering themselves primarily as philosophers, who tend to see themselves as outside observers of a specialised and privileged knowledge-generating activity.[6] The orientation to ontology that I shall be concerned with, and is currently receiving significant interest within the social sciences, appears to be quite different from this however. And although it has its roots in the same ontologically-oriented philosophy of science, it is better described as a collection of philosophically informed social scientists attempting to engage in ontology *for* social science (Latsis et al., 2007). Thus, there are quite marked differences in approach and emphasis.

First, there has not been a wholesale or straightforward adoption of philosophical terms, and where terms have been adopted from philosophical discourse, the meanings are often quite different. For example, for social scientists the term 'scientific realism' is intimately linked to discussions of ontology, and more specifically to discussion of the particular ontologies presupposed or implicit in the doing of different sciences, rather than the more epistemological preoccupation with discussions of truth and its reliability found in philosophical accounts. Secondly, social scientists perhaps inevitably have far more practical concerns. Thus social scientists tend to focus upon the nature of social items or categories such as social structure, social institutions, rules, conventions and norms, rather than warrants for knowledge, identity conditions, etc.[7] It is perhaps no surprise, then, that social scientists have tended to be more concerned with the work of philosophers such as Bhaskar and Searle, rather than Quine, Carnap or Putnam, for whom these latter categories are central. Moreover, the main attraction for social scientists has been the development of particular ontological conceptions rather than the treatment of epistemic problems raised by ontological explanations of the success of science.[8]

---

[6] This particular ontological orientation has recently generated considerable non-philosophical research. Perhaps the most significant example lies in artificial intelligence and the computer sciences, where 'formal ontology' is currently the central concept of a growing research programme. The aim is to construct formal representations of entities and relations in a given domain that can be shared across different contexts of application. At least part of the attraction appears to be the idea that 'philosophy offers rigour'. However, in information science the idea is not to model reality but to make it possible to integrate different data systems into a single or more encompassing system (see Smith, 2003). Philosophers have been teamed up with software specialists in the hope of providing all-encompassing top-level or 'backbone' ontologies that all data systems might be translated into.

[7] See for example Pratten (2014).

[8] It is worth noting that I am really only concerned with those developments that feature in the rest of the book, which have in fairness only really emerged in recent

A crucial further difference has been the kind of ontological strategy adopted. Whilst not all social science strategies have been the same, I wish to focus here upon one very prominent example which, following the work of Roy Bhaskar, involves the use of transcendental argument. Whereas Kant asks what human minds must be like for them to have the kinds of conceptions they do, Bhaskar suggests that we instead ask what the world must be like either for various conceptions to be prominent or for certain practices to be possible. In contrast to the projects noted above, certainly to those initiated by Wolffian deductivism, transcendental realist projects are thus always acknowledged to be fallible, interactive, historically sensitive, etc. And the aim tends to be one of beginning analysis with conceptions that are generally agreed upon, but important or significant for whatever reason, and then to ask what the world must be like for these conceptions to be as they are.[9]

Not only will resulting accounts be fallible, but they will be very much a product of their time, and so not 'out of history' or a priori in any sense. Before proceeding, it is helpful to provide some examples that illustrate this recent orientation to ontology. Moreover, the following examples discuss ideas that will be drawn upon in later chapters.

Experiment and Isolation in the Natural Sciences

As suggested above, in order to pursue an ontology of technology, a crucial ingredient will be some account of the nature of, and differences between, social and non-social reality. Although considerations of science once more provide a starting point, the strategy adopted is rather different from that of Quine. To anticipate, the kinds of ontological accounts I am interested in here are critical of any understanding of science that is based on a 'constant conjunctions' conception of laws.[10] The criticism of constant conjunctions conceptions also provides a good example of Bhaskar 's transcendental realist

---

years. Throughout much of the later stages of the twentieth century, most social scientists interested in philosophy were preoccupied with such issues as warranted knowledge etc., which arose more out of the work of Popper, Kuhn and Lakatos.

[9] It is probably also worth pointing out that the 'must' here is not of course a deterministic one. It is simply signalling the idea that if the world were a certain way, this would be sufficient to explain the phenomenon under investigation.

[10] See especially Harré and Madden, (1975), Secord (1986), Bhaskar (1978).

strategy.[11] The argument proceeds by asking what must be the case for experimental activity to be intelligible or to occupy the place that it does in prominent accounts of science.

A constant conjunctions conception, for Bhaskar, indicates the belief that there are regular patterns in events of the form 'whenever X then Y', and that laws are first and foremost a description of these regularities that are obtained in experiment.[12] Bhaskar then draws out the implications of two important observations concerning such a conception. First, most of the significant scientific results, in the form of constant conjunctions, only in fact occur in experimental conditions. Secondly, the results of knowledge obtained inside experimental activity are applied outside of it. In order to make these observations, and more generally to make experimental activity intelligible, Bhaskar argues, the objects investigated must be both isolatable and irreducible to events. It is important for what follows to explain these ideas in some detail.

Although many, such as Anscombe (1971) and Von Wright (1971), have noted that the sequences of events that result from experimental activity depends on human interventions, they failed to highlight the ontological 'distance' between the objects of investigation such as causal laws and the patterns of events that are constructed in order to investigate them. In other words, experimental scientists produce patterns of events, but these events are constructed meticulously in order to understand (provide an index of) what they do not produce. Those adhering to a constant conjunctions conception of causal laws appear to be committed logically at least to the idea that either people cause or change the laws of nature. Furthermore, given the fact that outside the experimental set-up such regularities do not hold, the question arises as to what might govern laws outside of the experimental set-up. It is not clear that those committed to a constant conjunctions view can say anything about such situations (or must argue that there are as yet no laws).

What, then, is the basis of causal laws? To make sense, the observations of experimental activity require that we acknowledge the

---

[11] It is worth pointing out that Bhaskar's focus upon experimental activity follows not from his implicit belief that experiments are essential to science or indicative of 'proper' science. Rather, it follows from the fact the all those disputing the nature of science accept a significant role for experimentation. However, Bhaskar argues, this agreement is belied by very little in the way of analysis of what experimental activity actually involves or presupposes about the objects of study in the experimental set up.

[12] Bhaskar attributes this position in the main to Hume.

existence of generative mechanisms that can operate outside of the closed conditions that enable their empirical identification. Thus, talk of causal laws requires a conception of causal agents, things endowed with causal powers. But the activity outside of the experimental set-up calls for more than causal powers. Rather we also need some conception of things in play. This is not simply a conception of something that would have had some effect in other circumstances (a counterfactual) but a transfactual, that is, something that really is in play irrespective of the empirical grounds for its identification.

In short, in order to overcome existing problems with the conceptualisation of experimental activity, it is necessary to invoke a complex or structured conception of reality in which experiences are not reducible to events, and events are not reducible to the underlying powers and mechanisms that generate them. On this account the experimental scientist must both trigger the mechanism in question and prevent interference of other mechanisms with the operation of the mechanism in question, in which case experimental results are able to cast light on the workings of different generative mechanisms that are not, themselves, reducible to the patterns of results observed.

This recording of a relationship between antecedent and consequent of a law-like statement is, of course, only a moment, although a significant one, in the process of providing an explanation. A full explanatory account will also typically involve a range of different cognitive materials, including analogy and metaphor, to construct a theory of a mechanism which, if it were to work as postulated, would account for the phenomenon in question. The mechanism suggested will then be subject to all kinds of empirical scrutiny in the context of competing explanations.

Central for arguments I shall make later, however, is the role of (the practical achievement of) isolating the working of some mechanism or set of mechanisms. For this to be possible, not only must reality be layered or structured, but mechanisms must actually be differentiable or isolatable to some degree. In other words, it must be possible to hold off countervailing mechanisms, and it must be the case that, much of the time, the mechanisms isolated do continue to behave the same way. A priori there is no obvious reason why this should be the case, but given the ex posteriori success of experimental sciences, then seemingly a good deal of actually existing isolatability seems undeniable. To see the significance of this point for the ideas that follow, I must first sketch out the account of social ontology that I draw upon in later chapters.

Social Ontology

Despite the recent prominent use of experiments of some kind in such disciplines as experimental and behavioural economics or psychology, it seems fair to say that experimentation, at least of the kind referred to above, plays only a minor role in social sciences and that the role it plays tends to be of a different kind to that outlined above.[13] If this is the case, why is this so, and what other strategies might there be for providing an ontological account of the social world? The answer to these questions is perhaps the central concern of social ontology.[14]

We can define 'the social' as the domain that necessarily depends for its existence upon the activities of human beings.[15] Thus a range of phenomena can be considered to be social, ranging from language and dance to tables and chairs. But my focus initially will be on those phenomena sometimes called core-social (T. Lawson, 2015) or cultural (Archer, 1995) that have a mode of existence that depends entirely upon the ongoing activity of human beings. In other words, whilst tables or chairs come into existence only because of the activities of human beings, they then continue to exist largely independently of social activity – their mode of existence for the most part is independent of human activity. Other phenomena, such as language, norms of polite-ness, the highway code, credit, etc., exist primarily through being con-tinually reproduced and transformed in and through the daily activities of human beings. Thus their mode of existence is, for the most part,

---

[13] There is little space here for defending this assertion here. However, for a more general account of the failure of modern economics to copy the successes of the experimental sciences see T. Lawson (1997). There is also no space to discuss the nature of experiments more generally, especially those arguments that cast doubt on the reliability and status of experiment in natural science (see e.g. Galliston). However, it seems fair to say that even amongst those such as Galliston, there is no argument that experiment plays the same role in social and non-social science.

[14] The particular social ontology I shall summarise here, and draw upon below, has many sources, including the early works of Roy Bhaskar (1989), the structuration theory of Tony Giddens (1984) and the morphogenetic approach of Margaret Archer(1995). But it draws mostly closely upon recent developments found within work by the Cambridge Social Ontology Group, see Pratten (2015).

[15] The use of the term 'necessarily' here is required to exclude phenomena that depend upon our actions in a contingent way, for example, if we have the technology to destroy the moon, but don't, we might otherwise be able to say that the existence of the moon is a matter for social ontology (T. Lawson, 2015).

dependent upon social activity, and so such phenomena are a core feature of social reality.

If social science lacks the relatively solid anchor point of reliable or replicatable controlled experiments, its various branches have the advantage that they deal with phenomena that are for the most part already conceptualised. Thus phenomena such as rules, relationships, languages, money, economies, technology, etc., are phenomena that have existing conceptualisations, even if these conceptualisations frequently need to be redrawn and re-articulated in the course of social scientific research. Moreover, human beings not only conceptualise the social world but also negotiate it fairly successfully on a day-by-day basis.

A central feature of capable human activity is that it draws upon a range of collective practices. These practices, such as driving on one particular side of road, wearing particular clothes in different situations, turn-taking in conversation, etc., may or may not be legally enforceable but they are collectively recognised by the grouping or community in which they operate. The term 'recognised' is intended to convey the fact that not all members of a community necessarily agree with, or are positive about, the merits of all collective practices, only that they recognise such practices to be the way that things are generally done in a particular context. The use of the term community is also intended to be very general; we are all members of many overlapping and often conflicting communities or groupings, but it is in relation to the meanings and accepted ways of doing things of these groupings that certain actions gain meaning and legitimacy. Thus such actions have differing and overlapping meanings in different contexts relative to the different communities within which they occur.

Another feature of social practices is that they embody routinised components; routine (predictable, enduring) behaviour seems to be a pervasive feature of the reproduction of such practices. Moreover, it is these features of endurability and widespread acceptance (or recognition) that allow collective practices to serve the purposes they do. So understood, and however collective practices might have come into being, they tend to coordinate behaviour through indications of how things are done and in so doing, provide a good degree of stability and even predictability.

In stipulating how things are done, collective practices also inevitably involve a normative component, stipulating how things *ought* to be done. Practices will endure if individuals conform to, and reproduce, them and there is some internal stipulation about how individuals ought to behave or ought to perform certain tasks.

Aligned to the normative aspect of practices is the idea that sets of rights and obligations are always a constitutive element of such practices. When driving in the United Kingdom, I have the right to park in certain places but not others and have a complex set of obligations to follow including the manner in which I treat other road users. Many of these obligations are explicit and legalised, but many are not, and many sets of obligations and rights are reproduced with little other than goodwill and trust in the actions of others.

However informal these rights and obligations may be, they can always be represented, or understood, in term of sets of rules. Such rules also underlie the routinised feature of social practices. Whilst such rules are an integral component of collective practices, they are ontologically distinct from them. Rules can be broken, as well as remain uncodified or unacknowledged. Thus they are not the same kind of thing as the sets of practices that ultimately reproduce or transform them.

It is also the case that certain practices are followed by some people but not by others; there is differential access to practices, and so the rights and obligations attached to different practices will also vary. Alternatively, the same point can be made by considering any community in terms of sets of nested or overlapping sub-communities with different rights and obligations. Thus the interactions of such groups as landlords and tenants, employer and employee and husband and wife although presupposing each other (are internally related) will for the most part involve access to different rights and obligations.

Acceptance of such features suggest one further category of importance, namely that of positions. It is clear that positions can either outlive their occupants or be transformed over an incumbent's lifetime. The rights and obligations relevant to different landlords and tenants, lecturers and students, husbands and wives, etc., are not dependent upon the occupancy of particular individuals. Indeed, it is better to say that rights and obligations are attached to positions rather than to people or practices, but that people and practices reproduce or transform such rights and obligations, and so positions. It is also the case that those occupying positions have different powers in virtue of the rights and obligations attached to the positions they occupy. In other words, power expresses the positioned rights and obligations to participate in specific others-affecting collective practices (T. Lawson, 2012). At a very general level, all forms of social being depend upon positions associated with some form of positional power. The term social relation is used to refer to accepted sets of rights and obligations holding between and connecting two or more positions; as rights and obligations involve some kind of power, then social relations are always power relations.

Although the foregoing provides only a very brief and schematic account of the social ontology I shall draw upon, it is possible to generate from it several conclusions that will be of central importance in the following account of technology. First, there is no reification of the general concept of social structure. There is no sense in which social material exists over and above, or operates behind the backs of, human agents. In particular, collective practices, and the (ontologically distinct) sets of rules, positions and powers that accompany them, have a very particular mode of existence – they are condition and consequence of action and are reproduced and transformed through action. This aspect is often captured in terms of what has come to be termed the transformational model of social activity (see Bhaskar, 1989); although structure is not reified as existing independently of social activity, neither is such structure to be interpreted voluntaristically as some epiphenomenon of action. Action could not take place but for the really existing conditions of action that this conception of social structure aims to capture.

Second, social material is best understood as emergent. Although the topic of emergence is particularly fraught in the mainstream philosophical literature, it has recently received a good deal of constructive attention in social science contributions.[16] In this context, emergence refers to a situation where new properties come into being at one level which, whilst rooted to some lower level, are not reducible to that level. A crucial role here is given to the novelty that arises from different forms of organisation (see especially T. Lawson, 2012). Rather than some mysterious process of transformation taking place, the emphasis is upon forms of organisation that generate new system-level properties that are neither independent of, nor reducible to, the components out of which the system is organised.

---

[16] Good examples are provided by Archer (1995), Elder-Vass (2010, 2007), T. Lawson (2012, 2013) and Pratten (2013).

It should be clear that the account of social reality outlined above provides a good example of such emergence. Collective practices give rise to emergent forms of organisation that facilitate different kinds of coordinated interaction, stability, etc., which would not be possible otherwise. In other words, emergent powers exist in virtue of the occupancy of positions not available without such practices.

Thirdly, social reality is not only processual in nature, but will most likely be internally related and irreducibly open. As noted, positions will tend to be internally related, and any attempts to isolate particular regions, mechanisms or aspects of social reality will tend to be at best only partially successful and, most often, be simply inappropriate.[17]

This last inference leads to one of the central propositions of the following chapters, namely, that the social and non-social world differ ontologically in that the causal powers that operate in each domain are not all equally differentiable or isolatable. To be clear, whereas the particular status of controlled experiments in natural science suggests that many of the important causal mechanisms of the non-social world are isolatable, this seems unlikely in the social world; not only is there little experimental success of the kind which seems to be taken for granted in natural science, but, given the account set out above, we have little reason for expecting such isolatability to be other than a rare feature of the social world. Indeed, process, openness and internal relatedness make such isolation of causal mechanisms rather unlikely.

Amongst other things, such a finding has implications for how we think about or attempt to explain different phenomena. Where isolation is possible, it makes sense to think in terms of relatively self-contained mechanisms that operate in pretty much the same ways under the same circumstances. Where it is not, the very process of thinking about the causal mechanisms of things makes rather different demands. It is never possible to think about each aspect of a whole of some complex situation. Some kind of simplification is required. However, there is a huge difference between thinking about a causal mechanism in such a way as to leave something temporarily out of view and in such a way as to assume that all other complicating factors do

---

[17] Such an account does not however imply that nothing worthy of the name 'science' is possible in the social domain. This would only be the case if the kind of controlled experiment referred to above is all, or even most, of what we mean by science. That this is not my understanding of science is discussed in the following chapter.

not exist. This distinction has been formalised in the contrast between the method of abstraction and the method of isolation (T. Lawson, 1992, Fullbrook, 2009, 203–230). This distinction proves to be particularly important for later discussions, especially in the chapters on autism and on Heidegger and Marx.

Latour and the Impossibility of 'the Social'

Although I shall draw heavily upon these distinctions between the social and non-social domains in the chapters that follow, it is fair to say that such a strategy is currently likely to find little sympathy, especially in much of the technology literature. Moreover, the account of the differences between the social and non-social offered above is set at a level of generality that many will find off-putting. Thus it may prove helpful to clarify how this distinction, as I understand it, relates to prominent contributions that tend to be critical of any kind of distinction between the social and non-social. Here I focus upon one particularly prominent example, that of actor network theory, and more specifically still, the work of Bruno Latour.

As noted in the first chapter, actor network theory, along with other constructivist approaches to the study of technology, has done much to clarify the different ways in which technology is irreducibly social. Actor network theory, however, differs from other constructivist approaches in the emphasis it places on studying all dimensions of technology and not just these more 'social' features; it is argued that non-social factors cannot conceivably be ignored or bracketed if any useful study of technology is to be undertaken. On the face of it, actor network theorists are operating with what might be termed a 'dual-nature' conception of technology, where the social and the non-social are combined.[18] However, this description is one that is unlikely to be

---

[18] Throughout such dual-natures discussions there is a degree of tension concerning what the 'other' for the social is. Is it material or natural stuff: material objects or nature? Moreover, of course, great scope exists for the manner in which the social and material might be combined. Often different uses of the terms 'function', 'form', 'structure', 'content', etc., are combined in different ways. For example, one currently very prominent account is provided by the Delft dual-natures approach (Kroes and Meijers, 2006, Kroes, 2010). Here a sharp distinction is drawn between the physical 'structure' of the technical object and the 'functions' that are ascribed to that object. Elsewhere, it

embraced by actor network theorists. Specifically, actor network theorists tend to argue that to conceive of technology or any other objects in a dual-natures manner involves acceptance of an analytical separation between the social and non-social that is undesirable and most likely impossible. Perhaps the most prominent advocate of this position is Bruno Latour.[19]

Latour seems to take emergence as an important feature of reality and argues for a particular ontology of the world as inhabited by agents with their own 'irreducible powers, capabilities, propensities, etc., interacting and remaking the world in a continuous process of interaction'.[20] Thus Latour embraces an ontology that seems very similar to that being defended here. However, Latour concludes that all kinds of irreducible actors, or actants,[21] come together in such a way that there is little point in distinguishing between social and non-social features. Latour is critical of both sides of what he characterises as the 'standard' conception of the social/non-social divide. For him, both the conception of the social and the non-social are problematic and act as sources of confusion in attempting to understand the world.[22] It is, however, 'the social' (especially as conceptualised in traditional sociology) that receives the bulk of Latour's attention.[23] The specific

is the form that is viewed as social and the content that is viewed as material or part of the natural domain, e.g. see Bhaskar (1986). In another account, it is the form that is material/structural and the function which is social (Faulkner and Runde, 2009).

[19] The main focus, here, is upon Latour's work since the publication of *Irreductions* (Latour, 1988).

[20] Latour explicitly argues that 'nothing can be reduced to anything else, nothing can be deduced from anything else, everything may be allied to everything else' (Latour, 1988, p. 163).

[21] A term used to signal that it is not only human actors that have causal capabilities.

[22] With respect to the latter, as noted the material is not something that can be ignored or bracketed-off. In particular, Latour is critical of those he terms 'social' constructivists, such as Collins, Bloor et al., who argue that there is no need for, or no possibility of, studying the material dimension of social technological phenomena. For Latour, this would be impossible. In truth, both sides of the debate tend to simplify each other's positions. However, it does seem fair to say that those such as Bloor tend to trivialise the role that objects make in any analysis (Latour, 1999).

[23] Latour particularly singles out the dominance of Durkheimian thought on the development of sociology and has spent much time in recent years rediscovering and popularising the work of Tarde (e.g. see Latour, 2009), understood as a major critic of Durkheim.

criticisms Latour makes of the resultant conceptions of the social are wide-ranging and evolve over time, but can roughly be summarised by pointing out some things that, according to Latour, the social is not.

 First, the social is not a zone or area (Latour, 1993). Here, Latour portrays modernity in terms of attempts to dissect the world into utterly opposed realms. On one side there is the human sphere, characterised by freedom of action, incommensurable perspectives, etc. On the other side is the external world, characterised by matters of fact, and precise, mechanistic events. Put differently, a sharp distinction is made between nature (the objective, material world), and the different perspectives that human agents have on it (see also Latour, 2004). For Latour, such distinctions make little sense; there are not two separate zones that would require the idea of 'pure states' of the social over here and the material over there that become combined somewhere else. There simply are no separate spheres that could be separated or linked or whatever. In contrast, reality is made up of all manner of acting objects that are interacting, conditioning and constraining each other. The real objects of studies for Latour are acting 'actants', sometimes called quasi-objects or hybrids. Despite the unfortunate use of such terms as hybrids, such objects are not 'mixtures' of different kinds of things, since 'pure forms' of the social and material do not exist.

 Latour's second line of criticism is that the social cannot be understood in terms of the working of 'social forces'. In talking of such forces, Latour is intending to capture such phenomena as power relations, inequalities, etc. The point can be illustrated in terms of his distinction between an intermediary and a mediator. An intermediary passes on the information or signal given to it without changing anything. For example, billiard balls and electric cables are perhaps most easily understood as intermediaries in that they simply transport some signal (force) without transformation. However, a mediator relays its information or force by amending it, transforming it, making it something else (an idea which is often captured by the term translation). In this case, the output of some situation is not well predicted from its inputs. The path along which some force is transmitted is littered with agents making a real contribution, transforming and recasting whatever is being transmitted. Thus, Latour argues, the idea of the social as a series of forces is problematic, given that in any case it is the individual actants using their particular powers and capabilities, transforming the 'signal' as they do so, that is important. The implication once more is that it is the details of the actants, linked and associated in particular ways that explains the phenomena under investigation.[24]

Finally, Latour argues that the social is not a kind of material (in particular, see Latour, 2005). Again the explicit target here is traditional sociology, where, Latour argues, the social is treated as a kind of adjective, such as wooden or metallic.[25] At the heart of Latour's concerns is the idea that once an explanation can be made in terms of social material there is nothing more for the researcher to do; explanation ends with an account in terms of social forces. In contrast, Latour suggests, we should always ask where such properties could come from, concluding that the social material 'answer' is nothing but a sleight of hand. Latour's alternative is to treat the 'stuff' of social explanation as 'gatherings' of mediating actants. It is not that social aggregates are given and so usable to shed light on those aspects of the world not dealt with or covered in other disciplines (such as economics, psychology, etc.,). Rather, social aggregates are to be explained in terms of specific associations of agents (including the phenomena usually distinguished as economic, psychological, etc.,).

In sum, Latour is arguing that it is impossible to distinguish the social and non-social. The argument is not that distinctions might be possible in principle but in practice prove impossible, given the complexity of the interrelations between the social and non-social (as, for example, Pickering's Mangle of Practice seems to suggest).[26] Rather, there is in principle no basis for such distinctions. There are no separate zones of the social and non-social, no social forces, no separate distinguishable factors in play, no social material. There are, instead, just 'flat happenings' where a range of actants cause and are caused, giving the social scientist little else to do other than trace (describe/collect) the nature of their associations.

The question I am concerned with here is whether the conception of the social that I am defending is susceptible to the sort of criticisms

---

[24] Although it could be argued that such an account of intermediaries works to distinguish the social and non-social, translation only really being an important part of the social domain.

[25] Rather strangely, Latour would include in this list of 'real' materials the economical and the biological (Latour, 2005, p. 1).

[26] We shall return to this idea in the context of a discussion of sociomateriality in Chapter 5.

that Latour launches at traditional sociology. I think it should be clear that it is not, but let me spell out why I believe this to be so. First, Latour is correct to assert there are no zones or areas that are social. The mode of existence of the social on the above account is through the actions of human individuals – there is no separate social domain or geographical location. To suppose otherwise is to make the mistake of attributing the same mode of existence to social and non-social phenomena. This is actually something that Latour elsewhere argues forcefully that we should not do (Latour, 2015). Moreover, the problem of treating the social as a kind of material is equally easy to set aside. The conception of emergence discussed above highlights the coming into being of new and different capabilities and powers through forms of organisation. But all such capabilities are themselves in need of further explanation (diachronically if not always synchronically). In other words, whilst reduction is not possible at a point in time, this does not mean that emergent powers cannot be explained over time (T. Lawson, 2012).

Lastly, what of Latour's idea that the social can have no forces or powers? Again, the conception of emergence noted above does seem to avoid the criticism. Certain systems are so organised as to have powers that did not exist before the organisation came into being. Societies are so organised that there are all kinds of pressures and obligations on those positioned in different ways. It is difficult to know how to describe these pressures as anything other than social powers. But these are collective powers that reside in the community or grouping involved. Again, the important point is that there is a different mode of existence of social phenomena. Comparisons to billiard balls and electricity cables thus seem rather beside the point. The idea of causal power here is used to convey the idea that there is pressure on those positioned to behave in a certain way; pressure to abide by obligations, rights that are due, etc. The kind of force invoked here is perhaps better captured by the term deontic power (Searle, 1995, 2010).

These pressures, however, do not actually have to be instantiated in particular actions for them to be real. For example, we take it for granted that rules can be broken, but for Latour such a statement would make no sense. If the rule exists, it is present in the 'event' or 'gathering' that the social scientists disassembles. Ultimately, Latour's criticisms only hold if the world conforms to his idea of flat happenings.[27] But if the world is structured, stratified and differentiated, then this opens the way for focusing upon kinds of mechanisms that operate without being actualised in Latour's sense. Latour's failure to break away from the kind of constant conjunctions ontology noted above seems to make this option impossible for him. If these arguments are correct, then the term 'social domain' does not refer to a place, or zone, but to something like a *set* of social mechanisms – mechanisms that operate and have any force only through the activities of individual agents.

Ontology and Definition

The aim of this chapter has been to introduce and articulate a particular conception of ontology. By locating this conception within a historical account of ontology itself and highlighting the difference between it and older notions, the aim has been to both clarify the conception of ontology with which I shall be concerned and also to perhaps ease concerns, which seem to dominate much of technology studies literature, that ontology is either without implication or deeply problematic.

At the same time, the aim has been to begin the process of setting out an ontological account of technology that combines or straddles the social and non-social worlds. The starting point for this has been an ontological distinction observed between the social and non-social world in terms of the relative isolatability of mechanisms that are causally significant in each domain. Here the general strategy of transcendental argument has been used (from experiment) to proceed to an account of what the natural world must be like.

Given the rarity of such experimental situations in the social world, a different approach is required. Of course, there is unlikely to be one best way to approach social reality and no one method that can substitute for controlled experiments. However, as noted above, there are compensations in the social world. First, much of social reality is already interpreted in some way, and secondly, there is rough consensus concerning all kinds of features of reality that can then be used as

---

[27] Although Latour does not, as far as I am aware, explicitly mention ontological accounts that are critical of event-ontologies, a close collaborator has more recently clarified the situation as follows: 'Latour is proudly guilty of what Bhaskar and DeLanda call "actualism"' (Harman, 2009).

premises for the kinds of transcendental arguments outlined above.[28] Such compensations of course do not suggest that it will be possible to arrive at the kind of law-like results of the natural sciences. But they do provide points of access and resources to be drawn upon in social-scientific explanation. They also underline the historically contingent, constantly reconfigurable nature of social explanation. Thus whilst it may be possible to proceed by transcendental argument to accounts of significant features of the world, this cannot be done in some ahistorical or timeless fashion as early Wolffian rationalists believed. Any ontological account will be conditioned by the historical context in which it is carried out. Moreover, social ontology is always dealing with preconceptualised categories and phenomena. Thus social ontology always involves a particular emphasis upon reviewing, and attempting to accommodate, existing conceptions and meanings of different phenomena relevant to any particular ontological inquiry.

To emphasise the partial, contingent, fallible nature of ontology is not to say that 'anything goes'. In this context it is useful to distinguish between two things that are often conflated, especially in some constructivist accounts, namely definition and ontology. How we collectively define things is clearly a matter of convention. How we decide upon units of measurement, which words are used to refer to particular animals, etc., is all conventional. Changing such definitions is not a matter of discovery but is one of persuasion. Ontology, on the other hand is not really conventional, at least not in the same way. There really are discoveries to be made about how things are in the world, what they are made up of and how they exist. Any account of them will always be fallible and partial, but a good account can be defended by more than reference to convention or tradition.

Ontology, however, still needs to be persuasive. And it seems hard to deny that the persuasiveness of some ontological account will depend a good deal on the selection of features to accommodate. The task of the previous chapter, which focused upon how and when the term 'technology' became an important term of social analysis, was to suggest features of technology that a significant and plausible ontological account should be able to accommodate. The task of the next five

---

[28] Indeed something very like this social ontology has been arrived at by transcendental argument from the routinised nature of social reality and the existence of collective practices (T. Lawson, 2015).

chapters is to accommodate such features into a significant and plausible ontology of technology, where the term 'ontology' is understood in terms of the conception of ontology advanced in this chapter.

In so doing, the following chapters engage in what is usually called scientific ontology, which can be contrasted with the philosophical ontology that has been the main focus of this chapter. To clarify, philosophical ontology deals with very general features of the kinds of things that exist, with properties and ways of existing of different phenomena. Scientific ontology can be thought of as a kind of applied philosophical ontology, in which particular phenomena found to be of significance are analysed in terms of the elements of being. It is to this latter task that I now turn, specifically focusing upon a conception of technology.

4 | *Science and Technology*

The relationship between science and technology is often viewed in a rather simple and clear-cut manner: science and technology are relatively discrete, easily distinguishable activities, the latter being wholly dependent upon the former. The idea that technology can be understood as 'applied science', dominant not only in academic contributions concerned with the nature of technology but as found in more practical documents, such as course guides for the study of particular technologies, is perhaps the most common expression of such a view. As noted in Chapter 2, the term 'applied science' was also central to the discourse in which both scientists and engineers sought to establish their status within the academy and wider society, and so important for the coming into being of the term 'technology'.

Conceptions of applied science, and the rather discrete separation between science and technology that they seem to imply, have been challenged in recent years by a variety of different contributors who call into question not only the starkness of the distinction but the direction of causation presupposed between technology and science. Although there is much of value in these criticisms, I shall argue for a somewhat different account of science and of the relationship between science and technology. In so doing, I seek to avoid the problems involved in holding a simple 'applied science' conception of technology, but also of conflating technology and science. Moreover, the account I want to propose also focuses attention on particular sources of similarity between technology and science which are rather different to those usually discussed, namely the special role occupied in each by the isolation of the causal properties of things.

Technology as Applied Science

Although the idea of applied science has a long history, it was not really formalised in a significant way until the contribution of Bunge

(1966).[1] Bunge suggests that the method and theories of science can be applied either to increasing the knowledge of reality or to enhancing our welfare and power: if the goal is purely cognitive, pure science is obtained; if primarily practical, applied science results (ibid., 329). Implicit in Bunge's position is the idea that technology results from science. Technology is about action, but that action is always underpinned by theory (to distinguish technology from crafts and other practical skill-based actions). This theory can then be divided into two parts, both of which can be seen as applied science, although in different ways. On the one hand, there is substantive theory, which is about the objects of action. On the other hand, there is operative theory, which is concerned with action itself (we might think of this now as organisation). Bunge argues that the former are clear applications of science. The latter are more complicated as they are not typically preceded by scientific theories. But, Bunge argues, they are dependent upon scientific method.[2]

Bunge's formulation has drawn a variety of different criticisms. First, the priority of science has been challenged. Some have argued that although science may have been dominant until recently, currently technology is the driving force of science (Forman, 2007). Others have argued that science has always been dependent upon technology. Anthropological evidence suggests that technologies, such as axemaking, use of wood, the construction of simple nets, etc., predate the

[1] As noted in Chapter 2, the terms 'science' and 'applied science' came into common usage in only the middle of the nineteenth century, and at this time technology tended to have an obscure and specialised meaning as the study of the useful arts. As also noted, from this period until the middle of the twentieth century, there was much vying for status on behalf of both engineers and scientists. Pure science was initially distinguished from applied science. But the idea that science might be 'pure' in some sense served both to paint scientists as irrelevant to the newly developing world of industrial machinery and revolutions in transport and communications. Moreover, engineers sought to establish themselves as part of a 'learned profession' rather than developers of simple craft or artisan activities. Thus, for scientists to be seen as important, to bolster requests for funding, and for engineers to distance themselves from untrained artisans, there was agreement between the representatives of both groups to carve up the disciplinary boundaries around distinctions first captured by pure and applied science, and then by science and technology (Kline, 1995).

[2] At about the same time, similar formulations were being considered. For example, Skolimowski (1966) made a very stark distinction between science as concerned with what is and technology as concerned with what is to be (almost exactly the same formulation can be found in the work of Herbert Simon (1969)).

human species, let alone human science (Ihde, 1983).[3] A related argument, common within the history of technology, is that the problems of technology have acted as spurs to important scientific breakthroughs. This idea lies at the heart of the familiar quip that 'science owes more to the steam engine than the steam engine owes to science', emphasising that the discovery of the laws of thermodynamics came from questions of energy loss in early steam engines, not from observations of nature (Gillespie, 1960). If correct, it would seem that the usually posited relation between science and technology may even need to be reversed (Vincenti, 1990, Constant, 1984).

A second set of criticisms of the 'technology as applied science' thesis have suggested that the terms 'science' and 'technology' refer to predominantly different practices and institutions which have had far less interaction than is often thought to be the case. For example, Project Hindsight, a study undertaken by the US Department of Defense to assess the impact of science upon military technology, found that science had relatively little impact: 8.7% compared with 91% of strictly technological sources (Sismondo, 2004, p. 75). Of course, implicit definitions concerning the exact nature of science and technology in such research are easy to challenge. But the basic point, that there are relatively self-contained institutions and practices which do not always cross over in any simple way, seems difficult to deny.

Third, many argue that technological knowledge is necessarily of a different kind to scientific knowledge, being more tacit (Laudan, 1984) or more empirical (Layton, 1974). For others, technological knowledge is often portrayed as a system of common sense (Vincenti, 1990). But whilst the tacit quality of technological ideas may have been relevant to some pre-industrial, community-based activities, it is hard to see such a sharp distinction in modern high-technology corporations or engineering companies.

---

[3] There does, however, seem to some disagreement about the details of relatively modern technologies. For example, whilst it is certainly the case that many technologies predate formal science, it has also been argued that the picture is complicated by a latent Eurocentrism in the historical accounts of technology. For example, as Francis Bacon observed, those inventions which most benefitted progress (but came about before a formalisation of science in the west, were paper making, gunpowder, the compass and moveable-type printing). However, all of these were actually invented by the Chinese (where a formalised 'science' seems more likely) (Bacon, 1623).

Whatever the postulated relationship between science and technology, most of the contributors noted so far have maintained a view of science and technology as clearly distinguishable phenomena. However, even this idea has recently come under concerted attack. These criticisms have come from a variety of different quarters, but share the notion that technology and science are so completely 'interwoven' as to make any real attempt at distinguishing the two, pointless or arbitrary.

A frequently made point here is that science has always required technological instruments to make any real progress. One obvious example is astronomy. Before the development of telescopes, when investigators had to rely on the naked eye, there was little that could be called a science of astronomy. But with the arrival of telescopes, astronomers could observe as never before. Furthermore, a series of leaps in astronomy can be linked to other technological developments, such as being able to use the full electromagnetic spectrum with radio astronomy. Moreover, the precise relationship between technology and science seems to change for different sciences. For example, contemporary biology uses a quite different array of instruments. We might even suggest that the biological sciences display a very different scientific culture, based on use of those instruments. For example, it seems that in biotechnology, instrumentalisation has encouraged a far more interventionist science, an obvious current example being gene-splicing (Fox-Keller, 2002).

The argument that accompanies such examples is that science and technology are so thoroughly interwoven as to make significant distinctions impossible (Ihde, 1991, Boon, 2004). Indeed, it is argued that it is impossible to talk of science or technology independently at all, and that it is much better to use the catch-all term 'technoscience' (Bruno Latour, 1987, Ihde, 1998). Using 'technoscience' in this way would, if taken seriously, seem to dissolve any discussion of the relation between science and technology. However, I believe that restricting discussions simply to 'technoscience' in this way, is both unhelpful and unnecessary. Rather, it is possible to distinguish science and technology in a way that avoids the above criticisms and tensions, but much depends, of course, upon what exactly we understand science and technology to be.

Experiment and Science

Much of the criticism of the 'technology as applied science' thesis has involved the idea that science has been reconceptualised in recent years.

The previously dominant positivist conception of science, with its preoccupation with mathematical forms of physics, emphasised the context-free and ahistorical nature of science. On this account of science, truth is gained more or less passively, using deductions and direct observations. However, such a conception of science has been significantly undermined by accounts that have been motivated by the practical observations of the actual activities and practices of scientists, especially of those in the laboratory. The activities surrounding controlled experiments have formed an important focus in these accounts, just because it would seem that it is in regard to experiment that the earlier positivistic conceptions of science are on their strongest ground. These critical accounts of positivism, which range from the social constructivist orientation of actor network theory and the strong programme to more realist interpretations, have emphasised that where scientific knowledge is obtained from performing experiments, such knowledge is the result not of passive viewing but of active intervening in the world, which itself involves the use of instruments. However, whereas the more constructivist critics as noted above have taken from this the idea that science and technology are not distinguishable, for those of a more realist orientation things are not so simple. Let me explain.

Consider for example the prominent account offered by Boon. According to Boon, 'science aims at creating phenomena by means of instruments and technological devices, as well as at the theoretical understanding of phenomena and instruments that create them' (Boon, 2015). It is this that is understood to generate the truly 'interwoven-ness' of technology and science. The emphasis here is upon the idea that the instruments of technology create or construct the content of scientific theories. If Boon's remarks are to be taken at face value, she seems to suggest a rather implausible form of constructivism. It is not simply that all scientific knowledge is the result of human activity and that the terms used in science are conventional, which realist contributors accept. Rather, if Boon is saying anything significant, it is that there is no discovery in science; experimental results are generated.

As noted in the previous chapter, however, the emphasis in more realist-oriented contributions is that discovery is central (especially see Hacking, 1983, Cartwright, 1999, Harré and Madden, 1975, Bhaskar, 1978). This is particularly clear in the work of Harré and Bhaskar. Both contributors emphasise the fact that experiment is an index of what the experimenter does not create. The details are different for different sciences, but in general, successful experimental situations depend upon isolating some particular mechanism or sets of mechanisms. Of course, the working of such mechanisms in the way observed is not something unique to experimental set-ups. To make use of such information, it is clear that such mechanisms must be operating outside the experimental set-up too, but in a way that is typically impossible or at least difficult to observe given the mass of countervailing and interfering tendencies and mechanisms. In this case, experiments are significant because they make it possible to hold off the effects of some mechanisms sufficiently to allow the identification of (the workings of) other mechanisms.

If this account of experiment is correct, how then should science be understood? It is notable that most critics of the 'technology as applied science' thesis spend little time explaining what science might be, other than to point out inadequacies in the positivistic notion of science as noted above.[4] Moreover, the literature most concerned with the importance of experiment has paid little attention to the workings, or potential 'scientific-ness', of disciplines in which experiment plays little role. Both sets of limitations are avoided if we adopt an account of science as causal explanation based on a concern with actual operating powers and mechanisms. Here, science can be understood to involve a wide-ranging set of activities. However, common to all these activities, is the movement in thought between events and states of affairs to the mechanisms and powers that generate them or underlie them but are typically out of phase with them. This movement is often termed 'retroduction', or following Peirce, 'abduction'. It is not primarily concerned with moving from the general to the particular (deduction) or from the particular to the general (induction). Rather, the central moment of science is understood to be a movement in thought from observations to powers and mechanisms which, were they to exist, would generate the phenomena under investigation. Making this movement in thought will require a range of different thought processes, such as use of analogy or metaphor, most often involving the transference of mechanisms familiar and understood in one context, discipline or perspective, to another. The importance and character of such movements in

---

[4] For example see Boon (2015), Ihde (2009) or Sismondo (2004).

thought will clearly change, also depending upon the role that replicatable experiments make in the particular science concerned.

On this conception, physics, chemistry, biology, economics and geography can all be understood to be sciences in the same sense, but they will differ in practice depending upon the nature of the causal mechanisms with which each discipline is concerned. In other words, and in keeping with the general realist orientation of this conception of science, how this retroductive moment is achieved or manifests itself will typically depend upon the object of study. Different object domains call for different methods.

Once the mechanisms and powers of interest have been 'identified', it is possible to distinguish an activity concerned more with understanding how the posited mechanisms and factors might combine to generate observed features of reality. This is what I shall refer to as applied science. If it is the case that the term 'science' refers to the movement in thought from events to causal mechanisms which could have produced them, and applied science refers to the use of such mechanisms to explain particular outcomes and states of affairs, we also require a term to capture the process in which mechanisms isolated in the scientific moment are actually recombined to produce new or different things with different or transformed capacities. It is this latter activity, I suggested above, that is best captured by the term 'technology'.

On this conception, technology is not applied science in the simple sense set out above, but it is closely related to some sciences. The situation is complicated by the multipart definition of technology I am adopting. Technology understood as study is certainly related to science, at least to those forms of science in which controlled experiments take a more central role. Given that experiments are most likely to be successful where the causal mechanisms under consideration are relatively more isolatable, isolatability also plays a crucial role in the possibilities for technology. However, there is no obvious relation of dependence in which technology is beholden for its theoretical breakthroughs upon science. Although technological artefacts may often be used to construct the kinds of test situations that are required for controlled experiments, this does not suggest either that technology and science, so conceived, are the same kinds of thing or that it makes sense to talk of a particular line of causation between the two. Indeed, none of the criticisms made of the 'technology as applied science' approach noted above, seem applicable to the understanding of the relation between science and technology that I am suggesting. It seems largely irrelevant which of the institutionalised versions of these activities came first, and seems quite likely that the actual practice of science and of technology are likely to involve different kinds of knowledge, different traditional approaches or bodies of knowledge depending on the problem at hand. There simply is not much that can be said at this level of generality.

On this account, there is no obvious need for a general hybrid term such as 'technoscience', unless perhaps to signify particular institutional contexts in which these different moments are so bound up together as to require a different distinguishing term. But there is no need for general replacement of both terms 'technology' and 'science' by a single term that is appropriate in all circumstances, because there are still different activities that the terms point out and articulate. Moreover, one central feature of technology, so understood, that a focus on science makes clear, is the moment in which causal mechanisms in the world are separated out not just in thought, but in reality, such that recombination of such mechanisms, and so the generation of new kinds of technological artefacts, is possible. Isolation and recombination are, then, likely to be essential components of any conception of technology.

Concluding Comments

In the account I am adopting, then, science is an explanatory exercise in which there is a movement in thought between observing particular states of affairs to the theories of causal mechanisms that generate them. In the more experimental sciences, this will typically involve practical attempts to hold off interfering or countervailing mechanisms in order to isolate and identify some relatively stable mechanism or set of mechanisms. Applied science can then be understood as the process by which existing knowledge of such mechanisms can be combined and organised so as to explain the occurrence of particular events or state of affairs. Understood in this way, neither of these activities is quite what the term technology is intended to capture. Rather, technology is concerned with using such knowledge of the world to produce objects that can be used for particular purposes. Thus technology, like much science, is concerned with the actual isolation and identification of different mechanisms; but technology is then concerned with their recombination to produce new kinds of things with different capacities and powers. The common denominator, as noted above, between technology and those forms of science most usually associated with it, is not simply the importance of experiment, but the possibility of isolation. It is isolatability that gives experimental activity its prominent place in the classic accounts of science. But it is also the isolatability of different causal mechanisms that makes recombination so important for technology. Thus this account maintains some kind of significant link between science and technology whilst also avoiding existing problems in the demarcation of the two. It also signals particularly significant elements or moments which will feature centrally in the ontology of technology that I want to defend.

There is, however, much more to technology than the isolation of causal properties of things and their recombination and much that can be said about the kinds of recombination that are possible. Indeed, this must be so if this conception of technology is going to be able to accommodate the kinds of features of technology outlined in Chapter 2. But it seems relatively easy to extend the understanding of technology so that such accommodation is possible. Rather than technology engaging in aimless isolation and recombination of causal mechanisms, I suggest that technology pursues such isolation and recombination in order to make it possible to harness the properties of things in order to extend human capabilities. Moreover, as noted in Chapter 2, the term 'technology' only really gains significance once its meaning is extended to cover reference to the material results of such processes of isolation and recombination. It is these 'results', in the form of technological artefacts, and how they are arrived at, that the following chapters are concerned with.

Specifically, each chapter asks one of the following questions about the nature of technological artefacts, questions which in one form or another have dominated the literature on artefacts, including technological artefacts. First, to what extent or how are technological artefacts social, not just because they are the result of human design, but in more ongoing ways? Second, what are the ontological features of technological artefacts that distinguish them from other kinds of artefact? Third, is there a predominant or characteristic way in which technological artefacts are used? In particular, how might we justify my contention that the use of technological artefacts is primarily to extend human capabilities? Chapter 8 is then an attempt to draw the threads of these earlier chapters together to articulate an ontological conception of technology that not only has significant advantages in its own right, but also manages to accommodate many of the significant contributions to the philosophy of technology noted in Chapter 1.

5 | *The Sociality of Artefacts*

The sense and degree to which technology can be understood as *social* is a point of significant tension in the existing literature. It was argued in Chapter 2 that the term 'technology' came into being to describe the largely material results of engineers and others in transforming the physical world. Thus the term gained currency as part of a shift away from the (social) activities of inventors and innovators, to the tangible (material, physical) results of such activities. On the other hand, it is often argued that one of the main strengths of recent (especially constructivist) theories of technology is their demonstration of the irreducibly *social* nature of technology (Brey, 1997, Bijker et al., 1987, MacKenzie and Wajcman, 1985b). In this context it is easy to see how the sense and degree to which technology is understood to be social has come to be something of a fault line in discussions of technology. This chapter is primarily intended to address these tensions.

Several points of clarification, however, are required from the start. First, the apparent tension here does not arise because of a distinction between technology and technological artefact; it is not that technology is an activity and so social, whereas technological artefacts are material and so not social. Rather it is the sociality of artefacts that is contested. Secondly, the statement that technological artefacts are irreducibly social may appear rather obvious. Artefacts are made by people and so, in a sense, must be social.[1] The more contested question, however, is whether or not, or in what ways, artefacts can be thought of as social in a more ongoing way, once they have been made. In other words, is there something about the ongoing mode of existence of artefacts that also depends upon the actions and interactions of human beings? This

---

[1] Certainly on the definition of the social advanced above, that the social is anything that depends necessarily upon the actions or interactions of human beings.

question has generated heated debate in recent years (Margolis and Laurence, 2007, Franssen et al., 2014)[2] and I shall draw upon some of this debate where appropriate. In particular, the main points of contention are set out by focusing upon issues of identification, function and practice, before suggesting some advantages of couching the discussion of artefacts, and their sociality, in terms of a conception of positioning as discussed in Chapter 3.

Two more definitions are required at this stage. First, in referring to the concern of this chapter as the sociality of artefacts, I am aware that the term 'sociality' has some quite precise meanings in certain disciplines. I am opting here for the more common dictionary meaning of sociality as referring to 'the state or quality of being social'. Secondly, and in anticipation of the following chapter, I am aware that much of the existing literature talks of objects rather than artefacts.[3] My reason for using the term 'artefact' rather than 'object' is simply to signal that I am most concerned, as I make clear in later chapters, with things that have been transformed to some degree by human activity.[4] The term artefact literally means to make by art, and for this reason users of the term artefact are typically emphasising the importance of makers in the coming into being of some object. My emphasis is especially upon the idea that things have been reorganised or reconfigured. I may pick

---

[2] Another important topic within such accounts is the extent to which different kinds of artefacts can be distinguished. This is the main concern of the following chapter, especially the need to distinguish technological artefacts as particular forms of artefacts, so I shall ignore this issue until then.

[3] However, I certainly do not want to downplay an interest in objects. For good reason, there is growing awareness that, within philosophy and social theory at least, objects have not received the attention they deserve. As noted, this is certainly a motivation for many contributions going under the heading of actor network theory. But more recently these interests have been taken up in a range of more explicitly philosophical contributions going under the headings of Speculative Realism and Object Oriented Philosophy. These developments have been explicitly realist and ontological in orientation (Bryant, 2011, 2012, 2014, Harman, 2011, 2013), arguing forcefully that objects have been essentially ignored since what might be termed 'the linguistic turn' (Ennis, 2011, Brassier et al., 2007). However, the particular ontological approach adopted has been significantly different from the one that I defend here, a point I shall return to later.

[4] It might be argued that use of the term object or artefact in practice hinges upon the significance that any particular author gives to the intentions of a designer in some particular account. However, to repeat, it is not the idea of a designer's intentions that I seek to highlight here, simply that artefacts are the result of some process of transformation.

up a stone and use it as a hammer or as a paperweight, and it may indeed be possible to consider both these activities as technological activities of some kind. However, I consider these to be rather uninteresting 'limit cases' given my goals of talking about technology's role in economic development and social change, and I shall use the term artefact to highlight this concern with objects that have been reorganised or transformed.

Artefact Kinds, Identity and Function

Much of the philosophical interest in artefacts revolves around issues of classification. A famous, early example is the ship of Theseus. The Athenians wanted to preserve the ship in which Theseus sailed to Crete to kill the Minotaur. They did this by replacing planks on the ship as they wore out. The question then arose, at what point would such replacement of planks produce a new boat, different from the one Theseus sailed to Crete? The problem can be taken further. What if the old planks are kept and at some point used to make another ship? Which of the existing ships has the better claim to be Theseus's? Are artefacts to be understood in terms of their component parts or in terms of their arrangement? More generally what determines an artefact's identity, what kind of thing is it and which particular instance of a kind is it?

With natural kinds, an important role tends to be given to a particular entity ceasing to exist. Particular humans, animals or plants die, and as far as we know this is the end of the matter. But the possibility of disassembly and reassembly makes artefacts quite different. As the example of Theseus's ship suggests, the attribution of identity is not at all straightforward. Is it simply up to us to decide which is which, or is there some truth of the matter that might be discovered? If it is the former, then there is something clearly very social about artefacts, which depends upon us in an ongoing way; if the latter, perhaps there is little that can be said at all about artefacts' sociality.

Attempts to address such questions, however, often end up addressing a slightly different question, namely, what is the metaphysical status of artefacts? For example, it is often suggested that the dependence of artefacts on human minds is unacceptable for an item's 'real' existence: 'if it exists, it should exist independently of what we care to think about it' (Franssen et al., 2014). Thus the sociality of artefacts comes from their 'mind-dependence' which serves not only to distinguish artefacts from such things as natural kinds, but in some way undermines their 'real-ness'; being social, in such accounts, seems to undermine the reality of things.

A related issue in the philosophy of artefacts is the relationship between artefacts and their functions. A concern with the example of Theseus's ship directs attention to the identity of a particular instance of a kind, a particular ship or statue or whatever. But can we also focus on the issue of distinguishing different kinds of artefacts? For example, we might use the same term, 'clock', to refer to radically different kinds of things (for example, a pendulum clock, digital watch and sun dial). It is in response to this apparent inconsistency that the functions of artefacts are often invoked as essential to identifying and understanding artefacts. Although constituted differently, certain artefacts can still be understood to be some form of clock because they are all used to tell the time; telling or recording time is what they are 'for'.

Many artefacts have names that spell out their particular function: lawnmowers are for mowing lawns, tin openers are for opening tins, coat hangers are for hanging coats, etc. In addition, it is in relation to an object's function that a range of other terms take their meaning, so that how an artefact performs in relation to its function is seen as the basis of its being valuable, good, worthwhile, broken, etc. Thus, an object's function amounts to more than a simple description of how something works, what it does or what it should be called. To talk of a bad tin opener or broken lawnmower is to talk of something that opens tins badly or that cannot mow a lawn, but it is still to talk about a particular kind of thing, a tin opener or a lawnmower.

As even these tin remarks indicate, of course, there are different conceptions of function at play. In particular, it is not always clear whether the term is being used as a noun, referring to a property of the thing itself, or as a verb, referring to the way that something operates. Thus the function of the bridge is to cross rivers, whilst the broken radio does not function correctly. Even if the term 'function' is restricted to its use as a noun, there is much disagreement amount whether or not the function is indeed a property of the thing itself or not. One particularly prominent account of functions, which adopts the position that functions are always observer relative, is to be found in the work of John Searle (1995). Searle argues that there is no function to be found in the object itself, and that the term only gains meaning relative to the interests and goals of human observers. Of particular interest here, Searle makes a distinction between agentive and status functions.[5] Agentive functions are ascribed where objects are used for particular purposes, such as when using hammers and screwdrivers. Within the class of agentive functions is a very peculiar, but for Searle very important, type of function that he terms status functions. The interesting property of status functions is that they ascribe functions to objects (and people) that they do not have solely in virtue of their intrinsic causal properties. Rather, the function refers to some capacity that the object or entity would not have but for a particular status it acquires in the process of being ascribed that function.[6] Examples of objects with status functions include cash, a driver's licence, identification cards, etc. Here as long as some group of people believe that an object has a particular function, this object does acquire certain (in Searle's terms, deontic) powers which it would not otherwise have.[7] Artefact functions, for the most part do not have such status functions attached to them. They are rather a special kind of agentive function which Searle sometimes terms causal agentive functions (Searle, 1995).

Searle is clearly more interested in the constitution of those aspects of the social world he terms 'institutional' rather than the sociality of artefacts. However, his account is important in several respects. First, for him, there are no separate things called social objects. The point is made most clearly in and exchange with Barry Smith, where Smith is critical of Searle's failure to include objects in his general social ontology.[8] Many phenomena, such as property relations, symphonies, laws, etc., Smith contends, are much more than concepts in people's heads, they are real objects. But for Searle, to call such objects *social*

---

[5] Although not important here, he also distinguishes non-agentive functions as functions pertaining to things that are not used for human purposes. In this sense, the heart, which has the function of pumping blood through the human body, is an example of a non-agentive function.

[6] Thus 'the President of the United States, a twenty dollar bill, and a professor in a university are all objects that are able to perform certain functions in virtue of the fact that they have a collectively recognised status which enables them to perform those functions in a way they could not do without the collective recognition of the status' (Searle, 2010, p. 7).

[7] For a similar position also see McLaughlin (2001).

[8] See Searle and Smith (2003).

makes little sense. His point is that a thing's 'social-ness' or sociality (whether or not, say, a five-pound note *counts as* cash) depends upon our descriptions of it. Thus, depending upon how we describe it, something can be both social and non-social. In other words, social-ness is a property of the way we view objects, and so there are no distinct objects called social objects.[9]

Searle's account is also important because although he has said little directly about artefacts, others have used his ideas to do just this by distinguishing technical objects as that subset of objects having agentive functions but not status functions. One important example is that of the Dual Natures Project. Here, although employing a slightly different conception of the term 'function' (as not simply observer relative, but as having a dual nature), technical objects are distinguished from social objects because the function that they have is 'performed on the basis of the object's own make-up' rather than on the basis of collective agreement (Kroes, 2012).[10]

Perhaps the central feature of dual-nature approaches is the idea that artefacts cannot be conceptualised either in terms of their physical characteristics or the intentions of their designers (or users) alone. Both must in some way be combined to give a coherent account of any artefact. In short, without some account of function, the artefact is simply a physical object, and without some account of the physicality of the artefact, there is no explanation of how or why something has a particular function. Thus the term 'function' serves as some kind of bridging concept between the physical and intentional conceptualisations (Vermaas and Houkes, 2006). But this then requires a 'hybrid' conception of function, which captures both the structural capacities of

---

[9] Or as Searle puts it:

> the notion of a social object seems to me at best misleading, because it suggests that there is a class of social objects as distinct from a class of non-social objects. But if you suppose that there are two classes of objects, social and non-social, you immediately get contradictions of the following sort: In my hand I hold an object. This one and the same object is both a piece of paper and a dollar bill. As a piece of paper it is a non-social object; as a dollar bill it is a social object. So which is it? The answer, of course, is that it is both. But to say that is to say that we do not have a separate class of objects that we can identify with the notion of social object. Rather, what we have to say is that something is a social object only under certain descriptions and not others, and then we are forced to ask the crucial question: What is it that these descriptions describe? (Searle 2003, 302).

[10] They also distinguish natural objects as those objects, such as electrons and meteorites, that have no functions at all.

the object, which are the concern of engineers, and the 'for-ness' of objects, how they are used and what they are used for (Kroes and Meijers, 2006, Meijers, 2009). In general, dual-nature accounts proceed by attempting to resolve a series of paradoxes or problems that arise in the ascription of functions to objects, and in particular how the identity of different artefacts is to be related to its function. A prominent example would be the malfunction problem. For example, if we understand a particular artefact to be a lawnmower because it can cut lawns, what happens when it no longer has this function? Is it still a lawnmower? In response to such questions, complex accounts of any object's identity are constructed combining different kinds of function.[11]

On the face of it, the different conceptions of function in the work of Searle and in the dual-natures approach would seem to make these accounts rather different. However, for present purposes the main point to emphasise is that there is actually very little concern given to the sociality of artefacts, as understood above. Rather, the focus is upon either the physical objects in themselves or the minds or intentions of those either designing or using such objects. For example, Kroes admits that he uses the terms 'intentional' and 'social' interchangeably; a use that is justified by suggesting that in either case it is intentions that are referred to, whether this be individual intentions or the intentions of groups (the social) (Kroes, 2012, 19). However, the conception of social I am interested in here, and as discussed in the previous chapter, is not reducible to intentions, or more generally to the mind, at all. Whilst the idea that groups have intentions in pretty much the same way as do individuals is highly problematic.[12]

I want to suggest that the main reason why both Searle's and the dual-natures accounts reduce the social to intentions, or to the human mind, is because there is little focus on human practices. As noted in the previous chapter, the basis for such phenomena as positions, relations, power, etc., is practice, especially collective practice. And it is to practice-based conceptions of the sociality of artefacts that I now turn.

---

[11] Prominent examples being the ICE theory of Houkes and Vermaas (Houkes and Vermaas, 2010, Vermaas and Houkes, 2006), the different kinds of proper functions suggested by Kroes (2012) as well as Preston's combination of proper and system function (Preston, 1998, 2003).

[12] See, for example, Lawson (1996).

Practice-Based Conceptions of the Sociality of Artefacts

Practice has featured strongly in a wide range of social theories in recent years (Schatzki et al., 2001). Practice theories are often portrayed as having their roots in the work of Wittgenstein and Heidegger (Shove et al., 2012). But it is clear that practice accounts of social reality are also to be found in many other places, such as in the work of Marx and the American Pragmatists (Rubinstein, 2013). However, a far greater coherence started to emerge in practice theories only in the late twentieth century. Obvious examples would be the structuration theory of Anthony Giddens, the transformational model of social activity of Roy Bhaskar, and the idea of bootstrapping to be found in the work of various post-Wittgensteinians such as David Bloor (Giddens, 1984, Bhaskar, 1989, Bloor, 1997).[13] Although these accounts differ in important respects, they all share several important features. First, and most obviously, set against cultural theories that locate the social either in the minds of humans, in chains of signs or symbols or in the intersubjectivity of speech acts, practice theories suggest that the social is situated, at least in part, in practice (see especially Reckwitz, 2002).

The second shared feature of practice accounts is the idea that action draws upon and reproduces pre-existing structural features of the social world. For Giddens, this is captured in the idea of recursivity; for Bloor this is bootstrapping; for critical realists this is at the heart of the transformational model of social activity. A central focus is upon the activity or performance in which, in the immediacy of acting, the social (however understood) is reproduced. Although there is some difference concerning the extent to which the social is reproduced and transformed in practice, or whether the social simply is the (fields of) practice (Schatzki, 1996), in the language of the previous chapter, social entities have a particular mode of existence that depends upon, but is not reducible to, human activity. Although artefacts have not figured centrally in any of the practice theories noted above, each of these forms of practice theory has been applied to some extent to the study of artefacts. And it is the differences between these various applications that are of most interest given current concerns.

Although Giddens writes almost nothing about artefacts, his ideas have been applied to their study in a variety of interesting ways.

---

[13] It would also be possible to include amongst the ranks of practice theories: Bourdieu (1977, 1990), Taylor (1971), Laclau and Mouffe (2001).

Giddens's approach (structuration theory), is often portrayed as the best known statement of the idea of recursivity,[14] that social structure is to be understood as both the condition and consequence of human action. However, it also has some well-documented limitations. Although social structure and human agency are presented as mutually constitutive (Giddens, 1984), there is actually considerable ambiguity about the exact ontological status of structure in Giddens' account. Giddens often talks about structure as generating a 'virtual order' of relations which are either only instantiated in practice or more often located only in the memory traces of human beings. But it is the latter that receives the bulk of Giddens' attention, and in later accounts Giddens emphasises that structure is only in agents' heads (Giddens and Pierson, 1998), thus often leading to criticisms of his work as subjectivist or voluntarist (Porpora, 1989) or of failing to emphasise the importance of the pre-existence of social structure for any particular action (Archer, 1995). In short, although Giddens presents a plausible mechanism for the way that social structure is implicated, in effect structure ends up all but disappearing from his account.

Perhaps unsurprisingly, a similar story can be told of much of the literature that has attempted to apply these ideas to the study of artefacts.[15] Authors have been attracted to the idea that technology and other artefacts result from the ongoing interaction of human actions and particular social histories and institutional contexts (Kling, 1991, Markus and Robey, 1988). But there is always an ontological priority that is given to the role of human agency in technical change and other interactions with technology (Orlikowski, 2010). In short, artefacts tend to disappear from the account much as structure disappears in structuration theory. This much seems to be accepted by many who have been prominent in initially applying Giddens' ideas to artefacts. One example is the work of Orlikowski, who has moved away from an account that explicitly draws upon structuration theory (the duality of technology approach) to one she describes as

---

[14] The term 'recursivity' is perhaps an unfortunate one given its different meaning in mathematics as a method of defining a sequence of objects, such as an expression, function, or set, where some number of initial objects are given and each successive object is defined in terms of the preceding objects (e.g. as with the Fibonacci series).

[15] Prominent examples are the adaptive structuration theory of De Sanctis and Poole (1994) and the duality of technology approach of Orlikowski (1992). For an excellent review see Jones and Karston (2008).

constitutive entanglement or entanglement in practice (2007). Here the idea is to develop Giddens's recursivity idea in a way that resembles Pickering's 'mangling' of the social and material in practice.[16] Such a position is described as a move away from a conception of a world of human beings and objects, each with separate properties (Law, 2004), to one in which it is impossible to know where humans and objects start and end (Introna, 2009, p. 26). The focus is then upon (system) capacities that are enacted or performed in practice (Suchman, 2007).

Whatever the merits of such an approach, especially in the context of remedying either an overly deterministic focus upon 'exogenous' technology, or the more voluntaristic approaches in which artefacts all but disappear, there is little help here with actually identifying the sociality of artefacts. Indeed, the project of articulating the social aspects of artefacts would presumably be seen, in line with the position of Latour outlined above, as rather pointless. But this focus does highlight a vagueness in these accounts. What is meant by entanglement? Entanglement of what? Who are the *we* who have such porous boundaries? It is indicative that the term emergence is considered central to much of this literature, even though Giddens was very critical of its use in any way to refer to the social (Giddens, 1984). If not articulated, the term 'emergence' is easily used to simply hide what is meant by the social. This would seem especially to be the case in these accounts.[17]

The second practice-based conception of artefacts I want to mention here has its roots in the work of Wittgenstein. Although clearly not as well developed a project as the dual-natures or structuration approaches, it has some definite advantages, given the focus of this chapter. As noted, in conceptualising the social, this approach prioritises practice above the mental (Bloor, 2001). Indeed practice is the 'primary generic social thing' (Schatzki, 1996). The central concept here is that of an institution. On this account the term 'institution' is used to refer to such phenomena as authority and money, which are distinguished as being: conventional, in that they are realised in different ways in different societies; self-referential, in that they

---

[16] There are clearly a range of different writers converging on similar ideas here, such as the relational ontology of Law (2004), and the sociomateriality of Barad (2014) and Suchman (2007), or the new materialism of Connolly (2013) or Bennett (2010).

[17] The general idea of entanglement, or mangling, of the social on non-material is returned to at the end of Chapter 9.

depend upon mutual or collective agreement, in a similar way to Searle's account of status functions; and performative, in that they are not determined by pre-existent rules but are continually brought into being through the actions of purposeful individuals (Barnes, 1988, Bloor, 1997). A central aspect of this approach is a feature that has come to be called 'bootstrapping' (see especially Barnes, 1983), which seems to be a version of Giddens' recursivity but where the role or formality of pre-existing rules are de-emphasised (Collins, 2001, Pleasants, 1996).

In addition to this conception of an institution, and the referential activity upon which it is based, there are three main conceptual and analytic categories – those of the natural, social and artificial kinds, which can themselves be understood as different kinds of institutions (Kusch, 1997). Natural kinds refer to such things as tigers or chemicals. Here the terms, such as 'tiger' or 'chemical', arise from social practice, but really do refer to existing objects (there is alter-reference), and a particular instance of the kind can be checked out or discovered by empirical examination. For social kinds, whilst the terms also arise out of social practice, the things they refer to are also the result of social practice, examples being authority or language. Artefact kinds are somewhere between these two other kinds. For instance, if we consider a typewriter, there is an object to which the term refers that exists independently of social practices. However, it would not be a particular kind of thing (typewriter) but for social practice. Therefore, whether or not some particular item fits the category, as with social kinds, depends upon social practice also.

The main implication of casting an account of artefacts this way is that it tends to de-emphasise the role of the designer's intentions in the understanding of some object as an artefact. For example, the dual natures project, which consciously attempts to re-establish the role of design and, what is often termed proper functions, is seen as misguided, or at least nothing much to do with the ontological status of artefacts (Schyfter, 2009). However, as defenders of the dual-natures approach point out, something about artefacts does seem to be lost in focusing so completely upon the social practices in which artefacts are used. More specifically, there seem to be a host of different examples, such as where the artefact is a one-off, where the artefact does not work well or where an artefact has a function prior to its use, that are not captured if the focus is restricted to the social practices in which the artefact is used

(Houkes et al., 2011). In general, it is difficult to see how this approach avoids the common criticism made of social constructivist approaches to technology that the materiality of the artefact ends up playing a rather impoverished role.[18]

It is against these problems of the recursivity and bootstrapping models that I turn to the form of practice theory that I argue is most fruitful in developing an account of artefacts and, as I make clear in the following chapter, of technological artefacts in particular.

Artefacts and Positions

A central aspect of the account of social ontology put forward in Chapter 3 is the idea that (core) social phenomena such as language, rules, credit, etc., have a mode of existence that depends upon their continuing reproduction and transformation through human activity. From the discussion provided in this chapter so far, it should be clear that artefacts also depend on human action, but depend on it in a variety of different, and apparently contradictory, ways. On the one hand, artefacts can be assembled, fashioned or modified by human action but continue to exist whether or not they continue to be acted upon by human beings; material artefacts such as hammers, paintings, houses, etc., once constructed, become relatively enduring independently of human action that brought them into being. On the other hand, such objects would not be 'hammers', 'paintings', 'houses', etc., if human beings suddenly ceased to exist. Pinning down the sociality of artefacts amounts to explaining these two apparently conflictual statements, that artefacts, once made, are both independent of and dependent upon human activities.

Elsewhere, I have couched the solution to this problem in terms of a transformational model of *technical* activity intended to compliment the transformational model of social activity familiar in critical realist accounts (Lawson, 2007, 2008, 2009, 2010). The formulation of ideas in these terms has been adopted primarily to draw out similarities and differences with regards to different kinds of activity. First, technical activity is simply a subset of all social activity, capably getting by in conditions not of our own choosing, involving, amongst other things, the reproduction and transformation of our conditions of action. But, and

---

[18] See Chapter 1 above.

this is the second point, the requirement for a separate model of *technical* activity stems from the fact that the consequences of some of our activities are often material artefacts which have a mode of existence that is relatively independent of that activity, and are thus quite different to the rules, relations and other elements of social structure with which the transformational model of social activity is concerned.

Using a hammer to bang in nails, for example, does not reproduce or transform the hammer in the same way that using language to issue a command reproduces language. Importantly, there is much that cannot be transformed at all. Gravity is not something that human beings can change, but something that must be discovered and then drawn upon. The importance of this will depend on the kind of artefact in question. Both a pendulum clock and a book are subject to gravity, but although a book may be very difficult to use in the absence of gravity, gravity for the pendulum clock is essential to its very functioning. But whatever the significance, there is a crucial sense in which using artefacts always involves *harnessing* the powers of existing mechanisms or affordances.[19] In short, technical activity is concerned primarily with discovering functional capabilities or affordances of different devices and mechanisms and drawing upon, or harnessing, such capabilities.

I shall adopt a slightly different emphasis here, however, and argue that the solution to the problem above (of how artefacts are both independent of but dependent upon social activity) can be understood in terms of the way that artefacts are positioned. Of particular concern here are the ways in which processes of reproduction and transformation are involved in the positioning of both artefacts and human beings.

To recap briefly, with regard to the positioning of human beings, capable human activity seems to presuppose the existence of a variety of collective practices, each of which establishes agreed ways of doing things. Linked to this is the idea that different people have access to different rights and duties in virtue of the positions that they occupy and accordingly enjoy different positional powers. The rules and routines that underlie or generate such positional differences are reproduced and transformed by the activity that is, at the same time, facilitated by the occupancy of such positions. As such these positions exist in relation to all kinds of different totalities or systems, particular

---

[19] This idea of harnessing is very close to the meaning developed by Pickering (1995).

communities, such as firms, families, etc., which in turn are understood as composed of positioned or organised individuals.

The question is then, to what extent can the positioning of artefacts be viewed in similar terms? On the face of it there seems to be significant similarities. For example, in building a house, different components are arranged or organised so as to facilitate the powers (of shelter, security, warmth, etc.,) that the built house affords. Thus artefacts, like people, are organised into totalities or systems within which certain powers and capacities are realised. Moreover, the transformational conception of social activity seems to be just as important in the use of artefacts, at least in so much as the positions which artefacts occupy are reproduced in a similar way through human activity. Thus, although a hammer is not like the core social phenomena discussed above in that its material existence does not depend upon human activity, the position it occupies (and its existence *qua* hammer) is reproduced and transformed through action in exactly the same way; it is its *positionality* that has the same kind of mode of existence as the core social phenomena discussed above.

But there are important differences in the way that people and artefacts are positioned. One important difference, relates to the role of rights and obligations. Whereas rights and obligations stipulate those things that incumbents of human positions can or cannot do, there is no exact analogue for rights and obligations in the case of artefacts. However, just as there are collectively agreed upon rights and duties attached to human positions, there do tend to be collectively agreed upon *uses* of, and procedures for using, different artefacts.

This idea has been formulated in terms of artefact functions. Specifically, positioning is understood to involve the idea that, in relation to wider systems, causal properties of the components are picked out that, in relation to some totality or system, can be understood as system functions (T. Lawson, 2012, p. 376). Thus, for artefacts it is functions that are attributed to the incumbents of positions rather than rights and duties as for human beings. For Tony Lawson, this attribution of functions also involves the artefact receiving an identity; when something is positioned as a bank note, or a hammer, that something receives a positional identity in much the same way that a lecturer or firefighter receives a positional identity.

Attaching identities to positions in this way seems to have several advantages. For example, the Ship of Theseus problem, noted above,

might be solved simply through suggesting that Theseus's ship is whatever the relevant community decides it is; if it is positioned by the
relevant community as the ship of Theseus, then it is the ship of
Theseus. However, to avoid the malfunction problem, identities (and
positions) must in some way be separable from functions. If we ask if
a broken lawnmower is still a lawnmower, the answer is that it is still
a lawnmower as long as it is so positioned, even if it no longer has the
function of being able to mow lawns.

I want to argue, however, that position and identity do not always go
together either, or at least not unless a very broad conception of positioning is adopted. For example, whilst it makes sense to talk of positioning
a pane of glass within a building's wall as so giving it the function of
a window, positioning a complex engine within a racing car does not seem
to work in the same way. Is it not also an engine whilst in the manufac-
turer's box? Thus can it be positioned outside the system in which it is
supposed to function? And so, at what point does it become positioned?
Is this once it receives a box with a label, or a safety check, a warrantee?
Crucially, can the artefact obtain its identity whilst still under construction or even at the design stage, and so prior to any use whatsoever?

I want to suggest that the only way to account for such examples is to
adopt a very wide and multidimensional notion of positioning that can
accommodate many very different practices. Rule-bound, collective
agreement about what something can be understood to be can then
compliment forms of positioning that involve the design of an artefact,
the reasons why it is reproduced over time, or the kind of Searlian
declarative act in which some agent (although on this account now
requiring the relevant positional powers, resources, etc.,) can simply
assert that something is a particular kind of thing.

An important factor here will of course be the extent to which the
artefact can be generally recognised as a particular kind of thing. If the
artefact is likely to be unfamiliar, a new invention or a dramatic adaption
of some existing kind of artefact, then the positioning of the artefact will,
at least initially, be determined largely by the designer or marketer.
The process whereby users come to hear about the artefact or how it is
supposed to work (advertising, review, word of mouth, instruction
manual, etc.,) will be particularly important here because they will
effectively set the terms for how some artefact is to be understood or
used. However, the Wittgensteinian point that most practices are entered
into by agents who already have some conception of different kinds of

artefact, is important here. Thus if we ask at what stage the engine becomes positioned as an engine, part of our account depends upon the pre-existent conceptions agents have of what is and what is not an engine. Then as soon as something is sufficiently constructed so as to resemble a car engine, the agreement that exists (about what constitutes a car engine) effectively gives it an identity as a car engine, even before it is positioned as an engine in any formal sense.[20]

Artefacts can be positioned in different ways, with different factors being more or less important in different contexts. Thus whilst engines and windows are pretty much mechanically slotted, and hammers and screwdrivers are used, food is consumed, paintings are admired, symphonies are listened to, video games are played, etc. Each seems to involve some kind of positioning in the above sense, but the processes involved are unlikely to be the same in each case. Indeed, there is probably very little at this level of generality that can be said about the nature of positioning of these artefacts, other than to provide the resources (a general social ontology) in terms of which individual accounts can be couched. Sometimes positioning seems to contribute much to the artefact's identity (panes of glass as windows, a traffic light) but not always (engines, photocopiers). At other times collective agreement in ascribing identity seems to contribute to the functions and uses that are viable (money, passports), but not at other times (hammers, screwdrivers). Sometimes there are important tensions between more and less formal aspects of such positioning (leading to important questions of whether an artefact is a real van Gogh or counterfeit money) and other times such tensions seem completely unimportant (is this a real clothes peg or toothpick?).

The point is that artefacts can be understood in terms of some account of their functions, positions and identities. Much of the time all three occur simultaneously, so that identifying capacities of an artefact as its functions simultaneously positions the artefact as a particular kind of thing. At other times, the link between the three terms appears not to hold at all. This much serves to underline the importance of, and an ongoing sensitivity towards, the particular ways in which different artefacts are positioned, a theme that is central to the concerns of the following chapter.

---

[20] Being formally positioned here might mean that it is given a quality assurance certificate, or something similar.

Concluding Remarks

To take stock, I am arguing that the sociality of artefacts comes, apart from the obvious sense in which they are constructed by human beings, from their being positioned, where the positions involved are reproduced and transformed through human practices. The sociality of artefacts then refers to those characteristics and features an artefact possesses in virtue of the way that it is positioned. For some artefacts, such features may be all important. For example, it is only because of the collective agreement that exists about cash or passports that they can be used in the ways that they are. For other artefacts, such as hammers and engines, positionality is only really responsible for their continued existence qua hammer and engine. Their continued use in practice serves only to maintain or reproduce their identities or the ways in which they are normally used. Their ability to perform a particular task, or to be used in particular ways, remains unaffected. Such distinctions are returned to in the following chapter. For now, the point to focus on is that the way that something is positioned remains central to an understanding of what that artefact is.

There are important questions about identity and positioning that I have ignored in simply suggesting that positioning takes place in different ways.[21] However, my main reason for not delving into such questions here is the belief that they are ultimately not so important for the kinds of artefacts I want to focus upon, namely technological artefacts. To say this, of course, I am committing myself to the idea that there is something about technological artefacts that accounts for a tendency to be positioned differently from other artefacts. This is the aim of the next chapter, to provide an ontological account of technological artefacts.

---

[21] In relation to similar accounts of the ontology of artefacts, such as T. Lawson (2012) and Faulkner and Runde (2013), my position would differ only in that I believe that what something is, is not exhausted by its positional identity. Thus communities can be wrong (for example, about whether some boat is the ship of Theseus, or whether a photocopier washed up on a desert island and used as a table, really is a table or a photocopier). Given my concerns here, however, I am not sure that any interesting implications follow from such differences.

6 | *Technological Artefacts*

The previous chapter emphasised the importance of positioning in understanding those features I am referring to as an artefact's sociality. Given this emphasis upon positioning, however, it may seem difficult to distinguish between different kinds of artefacts. If an important part of what an artefact is depends upon how it is positioned and used, distinctions between different kinds of artefacts would seem to reduce to how the relevant communities decide to use such artefacts, or to what such communities agree them to be. In which case, is there any basis for distinguishing technological artefacts or giving an ontological account of technological artefacts that might be different from other kinds of artefacts?

As my definition of technology stands so far, the term 'technological artefact' would roughly refer to the results of activities aimed at isolating causal properties of things that can then be recombined and harnessed to extend human capabilities. But such an understanding seems to undermine the importance of positioning developed in the previous chapter, in that being a technological artefact depends upon its history and not just how it is used.

This tension, between the importance of history and current use in determining what kind of thing some artefact is, has a long history in the philosophy of artefacts (an obvious example is the debates concerning the role and importance of proper and system functions) and still dominates the recent literature on the philosophy of artefacts, if often taking on new guises (Margolis and Laurence, 2007). In this chapter I shall again contextualise my position by reviewing prominent literature on this issue, but for the most part I shall focus upon contributions that are rarely considered in this context. Although the basis for distinguishing technological artefacts is not simple or clear-cut, I shall argue that there is much to say, ontologically, about technological artefacts that would seem impossible if being a technological artefact were to depend simply upon how something is used.[1]

Distinguishing Technological Artefacts

The term 'technological artefact' is most often used without much introduction or definition, seemingly in the belief that there is general agreement about which kinds of things are, and which are not, technological artefacts. Common examples include hammers, screwdrivers, computers, machines on production lines, aircraft, hi-fi equipment, mobile phones, batteries, drills, and so on. At the same time, such objects tend to be contrasted with other objects such as works of art, food, toys, passports, jewellery, etc., and although there may well be many objects at the margins, the belief still seems to be that there is a generally recognisable, and for the most part distinct, group of objects that the term 'technological' refers to (Lawson, 2009).

As noted in the previous chapter, accounts of artefacts are often bound up with discussions of functions. However, the term 'function' has an extra significance with respect to *technological* artefacts because such artefacts are thought to be 'for' practical, instrumental or ends-related purposes.[2] This is certainly the case in the dual-natures

---

[1] It is worth noting that in the literature there is little attempt to distinguish technical from technological artefacts. If the former derives from the term 'technical' and the latter from 'technological', there clearly is some difference (given the different roots of the terms technical and technological noted in Chapter 2). Here, however, I shall use the term 'technological' where possible, for the simple reason that I consider such artefacts to be part of technology. I also believe that most uses of the term 'technical' are motivated by similar concerns but that the term 'technical artefact' (or technical object) so dominates the literature that it is used often unthinkingly or at best inconsistently. Thus, unless I am referring to very specific uses of terms in the work of others I shall use only the term 'technological artefact', even if this is not always the term others have used.

[2] For some, distinguishing technological artefacts would seem to depend simply upon distinguishing technological functions, technological artefacts being those artefacts with technological functions. For example, Rathje and Schiffer (1982) drawing upon Binford (1962) distinguish techno-functions from socio and ideo-functions. Whereas the latter two kinds of function are symbolic, the former is strictly utilitarian, relating to such functions as storage, transport, alteration of materials, etc., (Schiffer, 1992). But the problem with using such an account to distinguish technical from other kinds of objects, is that any particular artefact may have all of these functions in different contexts. For example, a throne may have (1) the techno-function of allowing someone to be seated; (2) the

conception. Here technological artefacts are understood as artefacts that are material and serve practical purposes. There is an explicit reluctance to firmly distinguish technological from other kinds of artefacts, simply because it is very difficult to impose strict boundaries upon artefacts of different kinds (natural kinds, social kinds, etc.,). The strategy is rather to focus upon 'paradigmatic' examples of technological artefacts (Kroes, 2012, p. 13). But for present purposes, the main point to notice in this literature is that there is surprisingly little attempt to characterise technological artefacts in general, and that attempts to describe or analyse particular technological artefacts involve different combinations of functions, especially of system functions (how something is used in a particular context) and proper functions (which depend upon the intentions of its designer, evolutionary processes of selection, or whatever).

Elsewhere, technological artefacts are distinguished using Searlian conceptions of agentive and status functions. As noted above, agentive functions are ascribed where objects, such as hammers and screwdrivers, are used for particular purposes. Status functions are ascribed where collective agreement enables some artefact to have some power that it would not otherwise have. Although Searle does not use these distinctions to define a technological artefact, others have used them to do just this. For example, the contributions of Faulkner and Runde are notable in that they draw explicitly upon the kind of social ontology being defended here and are specifically concerned with technical object (Faulkner and Runde, 2009, 2013).[3] Drawing upon Searle's conception of observer dependent function, they suggest that an object is a technical object if it has an agentive function ascribed to it. Thus any object that is used for a purpose can be a technical object. One initial problem with this, at least in relation to the concerns of this book, is that, depending on how generally the term 'purpose' is understood, it is difficult to think of phenomena that would not count as

socio-function of communicating who is the king, conveying status, privilege, etc., and (3) the ideo-function of symbolising authority, monarchy, etc. Thus on the assumption that artefacts have technical as well as social and ideological (as well as aesthetic, moral, political, etc.,) dimensions, we at best have a typology in which many things can be viewed as technological artefacts in different contexts.

3 Faulkner and Runde explicitly adopt the term 'object' rather than artefact because they are concerned to generalise their account to what I above described as the limit case of objects used without any form of transformation by human. They also use the term 'technical' rather than 'technological' (see note 1).

technical objects. Indeed, Faulkner and Runde conclude that human beings can also be technical objects.

One way to proceed, using these ideas, might be to narrow a conception of technological artefacts to cover those artefacts that have agentive functions but not status functions. Or put another way, technological artefacts are those artefacts that do not require collective agreement or some conferred status for them to function in the desired way. This is fairly close to the meaning I shall give the term 'technological artefact'.[4] However, as it stands such a definition would still be too close to the idea that any artefact could be a technological artefact, if used or positioned in a particular way, and that there is relatively little, ontologically, that could be said of such artefacts.

Such a conclusion, I want to argue, is not quite correct. In order to suggest why this is the case, I want to approach the issues at hand from a somewhat different angle and ask what we can say about technological artefacts if we focus not on analytical examples of kinds or different conceptions of function but, rather, work backwards from accounts of technological change or the dynamics of technology more generally. Specifically, I want to focus upon the properties that technological artefacts require to account for the broad features of these accounts of technical change. In order to pursue this, I want to now focus upon contributions that are relatively little considered in this context, one set provided by the work of the American institutionalist economist, Clarence Ayres, and the other provided by the French philosopher of technology, Gilbert Simondon. Both, I argue, suggest ways of characterising technological artefacts that, in combination, turn out to have some interesting advantages.

Ayres, Tools and Icons

Ayres was one of the central contributors to American institutionalist economics. At the heart of this school of thought, or at least of Ayres's contributions to it, is a distinction often termed the Veblenian

---

[4] Dual-natures contributors also seem to settle on something like this distinction. Here differences between what they term social and technical objects are captured by the fact that the latter's function is performed on the basis of the object's own make-up, see Kroes (2012, p. 13). But there seems little discussion of the idea that settling upon such a definition might undermine the importance they attribute to proper functions in determining the identity of some artefact.

dichotomy or instrumental valuing.[5] Briefly put, a contrast is made between what are perceived to be the dynamic, change-inducing properties of technology and the constraining effects, or inertia, of social institutions. In various ways, good and progressive technology is counterposed to bad and obstructive social institutions. At the very least, technological change is seen as conducive to social change. More strongly, left unhindered, technological change continuously challenges the status quo, providing an irrepressible prod to existing social institutions to become more modern, appropriate and fair.

To contextualise a little, Ayres was attempting to escape from the kind of moral relativism he understood to be dominant in the social theory in his day.[6] For Ayres, technology embodied, and indeed exemplified, the instrumental logic of John Dewey and the life process of Thorstein Veblen, and ends up as Ayres's main focus of attention.[7]

---

[5] Many of course have argued that this dichotomy is the distinguishing feature of Institutionalism (see for example Waller, 1982, Munkirs, 1988).

[6] Unfortunately, there is little space available here to set this discussion of specific Ayresian ideas within an account of his general project and orientation. However, some very brief remarks may prove useful for the reader unfamiliar with his work. A central problem for Ayres is that if all judgements about what is good or bad are merely relative to the customs of a given community (a position that he felt was especially dominant in the anthropology of his day, for example in the work of William Sumner in 1906, and was beginning to 'corrupt' other social sciences) then the comments of any social scientists concerning the merits or demerits of a given policy are 'qualitatively indistinguishable from the mythmaking of savage society' (Ayres, 1978 [1944], p. 209). Ayres located the roots of this moral relativism in the work of British empiricists such as Locke and Hume, and in line with, for example, Russell, believed that such relativism was ultimately paralysing (Chalk, 1976). Ayres's response is not to search for a priori or indubitable standards against which different actions can be valued as good or bad (i.e. those ideas to which moral relativism appeared to be a reaction) but to try to locate value in particular types of actions, behaviour, outcomes, etc. It is in this regard that Ayres is usually understood to be drawing upon the work of Dewey and Veblen. Despite important differences between Veblen and Dewey, Ayres finds in the work of both authors, a similar concern with obstacles to processes that can be identified as essentially 'good'. Dewey's 'wall of privilege' and Veblen's 'ceremonial' or 'invidious behaviour' can each be seen as structural rigidities or obstacles, even though the 'good' processes they obstruct are different for each author (the dispassionate selection of habits of thought for Veblen and the process of self-realisation or education of the individual for Dewey). It is with this focus in mind that Ayres attempts to provide an alternative to moral relativism.

[7] It is left open here just how well Ayres's synthesis represents the spirit of either Dewey or Veblen's main contributions. For contrasting views see Mayhew (2010) and Hodgson (2004, Chapter 17).

Ayres spends a great deal of time arguing that technology is far more than the sum of physical or material objects.[8] At the very least, technology is understood to be a coming together of skills (technical activities) and tools (Ayres, 1978 [1944], p. 107). But it is clear that despite Ayres's general claims, it is tools that do the majority of the work in his account. Indeed, it is often this point that draws the most criticism of Ayres (see especially Strassmann, 1974, Brinkman, 1997). By way of justifying this focus, Ayres argues that it is tools that play the crucial role in progress. According to Ayres, to argue otherwise amounts to accepting that progress (such as moving from the Stone Age to the Industrial Revolution) is a result of greater skill or mastery of technique, a position that he concludes is completely untenable (1978 [1944], p. 112).

A central idea for Ayres is that all inventions are combinations of previously existing devices and ideas; inventions that appear to be spontaneous or discrete are usually the end product of a long series of combinations. Such inventions are not due to the 'magnitude of the soul of the "Gifted ones"' (1978 [1944], p. 115), mostly they are little more than the rather mechanical combination of existing devices, materials, instruments and techniques (1978 [1944], p. 113).

Ayres develops this idea, which he terms the principle of combination, in a variety of different ways. For example, he sees it as an important ingredient in understanding events that otherwise may appear as random or chance. For example, for Ayres '[t]he discovery of America was "accidental" with reference to the intentions of Columbus; but it was not accidental that it should have occurred in 1492. The arts of shipbuilding, seamanship and navigation being what they were by the end of the 15th century, somebody was bound to have "discovered America" within a decade or so' (1978 [1944], p. 115).

The principle of combination is also used to explain the frequent occurrence of 'simultaneous discoveries', where patents are often filed at close intervals in time (although not in space, and without any obvious communication of ideas between those filing the patents). Ayres argues that given the existing set of devices, materials, etc.,

---

[8] Ayres was clearly worried that the term technology 'suffers' from popular association with the most crudely mechanical techniques or with physical apparatus (Ayres 1978 [1944]: 155, 1978 [1961], 277–278). Ayres emphasised again and again that if the social dimension of technology was lost the significance of technology could not be understood (1978 [1944], p. 156).

somebody was bound (by putting them together in the right way) to make a particular discovery, whereas before such existing sets, that particular discovery would have been nearly impossible. Another example of the way that Ayres uses his principle of combination is to explain the huge rise in industrial progress in terms of the exponential growth or proliferation of technical devices: 'It follows that the more devices there are, the greater is the number of potential combinations' (1978 [1944], p. 119). This propensity of technical devices to proliferate, Ayres notes, is not a 'characteristic of men but of tools' (ibid.).

I shall return to Ayres' conception of recombination in a later chapter. Here, I want to focus in particular upon a distinction Ayres provides between tools and what he terms icons. Ayres repeats that all these arguments are based upon consideration of the tool-dimension of technology, and would not hold if technology were 'skills alone'. More precisely, Ayres argues that 'the things [that men of great ability – inventors] put together are physical objects' (1978 [1944], p. 115), and this is of fundamental importance to Ayres. However, it is not just the fact that tools are material objects that is important for Ayres. Icons are like tools in that they are material things used to accomplish certain ends. But the effectiveness of a tool, in contrast to an icon, does not depend upon the social status, standing or relationships of the tool user: 'a tool is an artefact which will perform to much the same effect whoever wields it, one that anybody can employ. A fetish [or icon], on the other hand, is wholly ineffective in any but consecrated hands. Profane hands may whirl the bullroarer in defiance of taboo and may produce a noise; but that noise will not summon any spirits' (1978 [1961], p. 135).

For Ayres, tools, in contrast to icons, have transcultural characteristics. And for this to be so, as well as the (exponentially increasing) possibilities for recombination, simultaneous assembly in different places, etc., they must have components that are relatively isolatable and can travel, and this is their physicality or materiality. But of course, icons also have a physicality or materiality. Thus it is not simply physicality that is important for Ayres. Rather, the important point is that, in the case of tools, it is their physicality or materiality that most immediately explains their use in particular ways.

Elsewhere I have recast Ayres's argument by drawing upon recent distinctions between social and technical objects, arguing that using social objects primarily involves harnessing extrinsic properties and

using technical objects primarily involves harnessing intrinsic properties (Lawson, 2009, 2015). For example, when we use a hammer to
mend a shutter we are primarily using powers *intrinsic* to a hammer to
perform the task at hand, such as its weight, the proportions of its
handle to its head, and so on. Two points are important here. First, the
use of the hammer does not depend upon the way that its user is
positioned. It may be that using a particular hammer in a particular
way may indicate the user is a carpenter or a plumber. But such social
positions, and their possible reproduction in use, are not required for
a hammer to be used. Secondly, a point Ayres makes forcefully again
and again, a hammer is a tool rather than an icon because the properties
of the hammer that we are harnessing, such as the hardness of iron and
steel relative to wood or plastic, are pretty much the same in all cultures
and in all times. Compare this to the use of, say, a passport. The main
causal power of a passport becomes obvious to anyone who has forgotten to take it to the airport when travelling. Without a passport, it is
impossible to board an international flight or travel across international borders. The power of the passport to enable its bearer to travel
between countries is inherently relational in character. It depends upon
a whole network of relations that exist between the bearer and the
passport, between the bearer and the airport staff, between the bearer
and his or her own nation state, between the nation states that the
bearer is trying to travel between and so forth. These relations, themselves, depend on a whole network of positions and practices.
As different materials come and go, and some technologies for identification become obsolete, it is the relational properties of the passport
that endure over time.

It might be argued that this emphasis upon the intrinsic suffers from
two related problems. The first is that technological artefacts can often
be described or understood as harnessing things that are extrinsic too.
For example, when I use a sailboat to travel, I am using it to 'harness' the
power of the wind. In other words, I am using intrinsic features (the
strength of the mast, the lack of holes in the sail, the waterproof hull of
the boat, etc.,) to harness powers that are extrinsic to the boat itself (such
as wind, reduced friction involved in travelling through water, etc.,).

The second, related, point is that it is possible to argue that all use of
objects, of whatever kind, involves the harnessing of intrinsic properties. Thus the priority of harnessing intrinsic properties when using
some artefact cannot be used as a way of distinguishing between tools

and icons.[9] From this perspective, it is not the passport that has the power to facilitate my travel between countries. Rather it is the deontic power afforded by my citizenship that allows me to gain access to other countries – the intrinsic capacities of the passport simply serve to identify me as a citizen with relevant rights. As such, attempts to distinguish between uses of such artefacts as hammers and passports cannot be based upon the relative importance of intrinsic properties in their use, because intrinsic properties are involved in the use of each artefact.

The above objections to using the term intrinsic to distinguish technological artefacts directs attention to the fact that the use of artefacts usually involves inserting them into, often complex, systems of interdependencies in which both intrinsic and extrinsic capacities are drawn upon or harnessed. However, I want to suggest that even if this is admitted, Ayres's distinction is still important. Specifically, Ayres's distinction focuses attention upon the different kinds of causal powers involved in the use of different artefacts. With the passport, as noted, the way that the user of the passport is positioned is essential to the passport's use. It is in virtue of the citizenship of the passport's owner, that travel between countries is permitted. The passport simply identifies the owner as possessing the relevant rights and obligations. Thus the power that the passport is able to harness is the *deontic* powers that results from citizenship etc. Such types of power are more or less irrelevant in the case of the hammer or the sailboat.

If this is the case, I think that the distinction Ayres is interested in can be recast as follows. Although the intrinsic features of an artefact are used or drawn upon in positioning it, some systems involve (require) far more for the artefact to be used effectively. In fact, in systems that contain people as important components, the coherence of the system itself is provided by all kinds of social phenomena such as collective agreement, social relationships, trust, etc., which are not intrinsic to the artefact. In which case, particular artefacts in the system will only be useable if a range of social phenomena operate. For example, collective agreement may be involved both in the way that the artefact comes to be positioned (there are likely to be legitimate or collectively recognised ways in which passports can be obtained or some artefact can serve as

---

[9] Although concerned with different matters, this seems to be the main point of difference between Searle and T. Lawson (2016).

a legitimate form of identification) but also in maintaining the system itself (in the case of crossing international boundaries, the very existence of such boundaries of course depend importantly upon such agreement).

For the artefacts Ayres identifies as tools, neither of these considerations appears to be of much significance. It may be the case that collective agreement will play some role in deciding what kind of thing an artefact is, or which tools are conventionally appropriate for a particular job. But typically such agreement will not really play much of a role in the use of that artefact. If a hammer is useful for banging in nails, then it will be so wherever and whoever is holding it, even if it is not the conventionally acceptable tool to use. In other words, the system within which the hammer is used (we can describe this in many ways, such as extending the swinging arm to bring increased force to bear on a particular point, etc.,) will tend not to have much role for collective agreement, social relations, etc., or what Ayres calls institutional features.[10]

When Ayres talks of anthropological artefacts (artefacts brought back from different cultures by anthropologists) as leaving much behind, he is referring to the impossibility of bringing back both the institutional factors that facilitate their particular positioning in different ways in different cultures and also to the role that institutional factors play in the systems in which such artefacts are used.

Viewed in this way, it is possible to understand why Ayres links the use of tools with change and progress and the use of icons with inertia and vested interest. Use of icons to perform some task will tend to (typically unintentionally) reproduce the sets of collective agreements, practices and positions required for that artefact to work. For tools, however, particular tasks can be performed without reproducing such relationality. Moreover, for reasons focused upon in the Chapter 6, new tools will tend to disrupt existing networks of relationships and positions. Whether or not this is an unambiguously good thing, as Ayres contends, is an issue that we shall return to in later chapters, but the impulse for change that Ayres outlines seems plausible and important.

---

[10] It may be the case that in a particular job, somebody is hammering in a nail to convince another that they are a trained carpenter, and thus such agreed ways of doing things may have more bearing, but this hardly seems an important exception to the distinction Ayres is trying to make.

Ayres's ideas about the transculturality of tools can then be summarised as follows. In using icons for some particular purpose, the user becomes effectively linked into a wide set of social relationships, that are cultural in nature and in which the user's own occupancy of particular positions is likely to be important. In using tools, such linking is unimportant. Thus tools work pretty much the same way wherever they are used. Whereas using icons tends to reproduce existing social and institutional relationships, using tools does not.

Even if this recasting of Ayres's argument is accepted, however, it still does not move us much further forward with the impasse identified earlier – to what extent do such distinctions really capture anything other than differences in the ways in which artefacts are used? Perhaps defining technology requires no further development of this distinction. An interesting recent example of a position that might suggest this is provided by Soltanzadeh, who suggests that technological artefacts are best understood as material objects whose physical properties are used to solve particular problems or puzzles (Soltanzadeh, 2015). This at least captures some of Ayres tools-icon use, but makes further distinction unnecessary. On this definition, of course, anything could be used as a technological artefact, in principle. It is only once we use something to solve a puzzle (we position it in a particular way) that it becomes a technological artefact (Soltanzadeh, 2016).

It is clear, however, that Ayres believes there to be more going on here than different uses or different kinds of positioning. For one thing, such an idea seems entirely counter to his account of progress involving the proliferation of new devices via a process of recombination. How can the kind of rapid or exponential developments he focuses upon be understood simply in terms of the differential use of ostensibly similar kinds of artefacts?

There would seem to be two ways to maintain the distinction Ayres is interested in. The first is effectively to argue that tools cannot, because of the kinds of thing they are, be used as icons; tools cannot be used to link the user to networks of social relations, collective agreements, etc., and do not depend upon the positional capacities of the user.

Perhaps more interestingly, however, the distinction can also be maintained by focusing upon the complexity or the structural integration of technological artefacts. Consider using a photocopier. The most striking feature of a photocopier is its physical or constitutional complexity but functional simplicity. Many different parts all come together to do one fairly obvious thing; paper and ink is put in at one end and returned, as a copy, at the other. There may come a time when archaeologists are uncovering the remains of our current civilisation and working out what all our artefacts are for. A passport may be subject to a range of interpretations (because as Ayres notes, much is left behind), the photocopier (if one survives intact) will more likely not. It should be pretty clear what a photocopier is for. This is because its intrinsic capacities, or physical make-up, tell us so much of the story. Of course, to be 'functional' is must always be positioned. It must be used by people who know how to use it, be plugged into an electrical system, etc. And it can always be used for other things – it could be sat on, used to hide a hole in the wall, etc. But it is still possible to work out what that artefact was designed for by investigating the structure of the artefact itself, with little recourse to the system within which it is used. Thus the photocopier is a very clear example of a tool rather than an icon, in Ayres's terms, because we can work out what kind of thing it is used for by a process of reverse engineering.

It is this tendency towards integration that, it seems to me, explains some of the intuitions discussed above concerning the manner in which technological artefacts often seem to have identities prior to any form of positioning or enrolment. The structural integration of the components of a car engine or photocopier is so 'tight', that it is almost impossible to imagine that they might be used for anything else. Of course engines could be used as paperweights and photocopiers as tables, but such use seems to be missing something. If reverse engineering is engaged in, in any kind of serious manner, and attention is paid to the knowledge gained from reverse engineering of these devices, it is hard not to conclude that they were used as engines and photocopiers.

Does this mean that we have a hard and fast distinction between technological and social artefacts? Not yet, even though it does seem to account for what Kroes and others term 'paradigmatic cases' of technological artefacts (Kroes, 2013). To say more, we need to argue that there is a tendency for technological artefacts to be more structurally integrated.

Simondon and Concretisation

In this regard, it is helpful to consider the work of Gilbert Simondon (1964, 1980). Simondon distinguishes sharply between the usefulness of a particular technology and what he calls its technicity. Technicity is essentially that which distinguishes the technical from other things, something which is established apart from its usefulness or function. In brief, for Simondon, the technical has a particular *form* of development. Central to this kind of development is the process whereby devices and technological artefacts come to incorporate more and more functions.[11]

Simondon's work is packed with specific examples given to support this idea. For example, air-cooled engines combine the functionality of a working engine with the ability to cool itself, thus no longer requiring a separate cooler. A rifle is structured so that it is able to use its own recoil to position the next bullet ready for firing, and so on. Simondon terms this process 'concretisation' and as particular artefacts become more structurally integrated they become, in Simondon's terms, more concrete.

Simondon's account has several advantages. Specifically, it incorporates a notion of growing sophistication alongside the idea that technologies come to fit or accommodate their social and natural environments. Thus the uses to which different artefacts are put end up becoming a factor in the kinds of functionality that becomes internalised or encoded into the artefact itself. In this way, it is possible to retain Ayres's concept of growth in knowledge without his overly deterministic conception of 'progress'.

Another strength of Simondon's account is that, in the context of Ayres's ideas about recombination, he gives us reasons for believing that the structural integration of technological artefacts will tend to increase over time and so their positioning will tend to differ in general terms from the positioning of other objects. It is not the observer relative functions ascribed to artefacts by people, but the internal or intrinsic functionality that indicates which kinds of artefacts are likely to be technological artefacts.

To be clear, combining the ideas of Ayres and Simondon, might not give us a hard and fast definition that can clearly demarcate between technological artefacts and other kinds of artefact. But, and this proves

---

[11] It is worth pointing out here that Simondon is using a rather different conception of function than that discussed above. When Simondon uses the term he is drawing upon the engineer's rather than the biologist's typical use of the term. Thus he is focusing upon the idea of working capacity or 'functioning', rather than purpose or 'for-ness' (Simondon, 1980).

to be useful in later chapters, it does seem possible to say a fair bit about the ontology of those artefacts that seem most deeply implicated in the kinds of technological innovation and progress that concerned those such as Ayres. Technological artefacts, on this account, are relatively isolatable, material objects, which tend to have tight, structurally integrated components and do not rely significantly upon collective agreement or social relationships to be used effectively.

Emphasising the 'becoming' or development of technological artefacts here takes us back to our initial distinction between artefacts and objects. In the light of the position reached, it should now be clear why a focus upon technological artefacts rather than objects has been adopted. Ultimately it is in the ways that things become transformed and organised that the technological takes on its distinctive characteristics.

Technological Artefacts and Technology

Although there seem to be significant advantages to combining elements of Ayres and Simondon's account, neither have much to say about the issues of sociality with which the previous chapter was concerned. However, it seems quite possible to supplement the ideas of Ayres and Simondon with something like the account of positioning and the account of science of the previous chapters. Ayres's main contribution is to argue that for tool-use to be effective, such use requires harnessing causal powers that are intrinsic to the artefact itself. Using a hammer to mend a shutter may reproduce the artefact-position 'hammer', or it may signal that the person doing the mending is a fully trained carpenter, but neither are necessary to actually mending the shutter. By contrast, we are harnessing the deontic properties of nation states, citizenship rights, etc., when we use such objects as passports. In other words, what Ayres terms icon-use (using artefacts that require larger systems of social relationships etc., to be useful) facilitates access to structures and relations. Their intrinsic properties, which act to signal the relations or positions of users or bearers are only of use or of importance in allowing access to a set of relations that are the properties of communities and other groupings, not of individual artefacts. Put another way, technological artefacts are relatively isolatable, in the sense that they can be used without the operation of wider societal features that are required for such objects as passports to be useable.[12]

Technology and Materiality

Given the emphasis in the above upon material artefacts, and on the materiality of technology, it may appear that my account misses an important, if not currently *the* most important, form of technology, namely, software or computer code. It seems undeniable that computer-based technologies have come to take a central role in modern societies. From opening doors and the operation of traffic lights, to authorisation of credit cards and the recording of data on just about everything we do, code and the processing and computation required for it to function in a variety of different devices, has become essential to how our world functions. And whilst there has been a wealth of analysis of software and coding and its role in modern societies, code itself remains a deeply anomalous or ambiguous entity. It is ambiguous in the sense that it tends to be thought of both as potent technology but also as something ephemeral, as non-material objects (Faulkner and Runde, 2016). Elsewhere this tension is captured in terms of 'the ambivalent ontology of digital artefacts' (Kallinikos et al., 2010), or digital artefacts are described as 'quasi-objects' (Ekbia, 2009) or as simply 'dubious' (Allison et al., 2005).

---

[12] I have laboured this point for two reasons. First, those still concerned with Ayres's work have tended to soften Ayres's distinction in the belief that contradictions in Ayres's general position suggest that technological artefacts, or tools, are irreducibly social and so must include a range of social factors. However, such a strategy seems to undermine Ayres's whole contribution. Rather than soften the distinction between tools and icons, I want to suggest that the most useful way to proceed is to maintain the distinction as part of a more general conception of technology where the sociality that critics find lacking in Ayres's account, can be supplied by ideas of positionality on the lines suggested in the previous chapter.

The other reason for labouring the last set of distinctions is to draw out Ayres's concern with the isolatability of artefacts. Indeed, isolatability is the primary feature of technological artefacts (or tools) that drives Ayres conception of technology. Although never explicit, Ayres is ultimately concerned with the differences in isolatability of different artefacts. This idea, of differential isolatability is one that is returned to throughout later chapters.

In the above, I have suggested that there is something very important about the materiality of technological artefacts. However, this does not mean that I believe software and code to be irrelevant to a discussion of technology. Rather, I want to argue that code and software are important examples of technology, but only because, or where, they are realised in particular material devices and artefacts. It is certainly possible to imagine code as disembodied, as ephemeral, etc. But to the extent that such things can ever exist as disembodied in any real sense, I argue, they are not technology. Moreover, I believe that this position is not actually as counterintuitive as it may at first seem. Let me explain.

The distinctiveness of computer software can be approached from a range of different perspectives. Often, at least with economists, the emphasis is upon the knowledge-based character of software as products or commodities. In fact, software or code is usually understood simply as 'pure' knowledge. So, for example, software is often described as non-rivalrous, non-perishable, etc.[13] The point is that we cannot trade knowledge as we do, say, most material commodities. Whereas cabbages can be exchanged for cash or some equivalent item, typically the parties to the exchange do not walk away with both items; one is given up for the other. However, knowledge can only be shared. If one person tells another something, then they both know it. Software is typically thought to be analogous to the latter situation. This is because once some code is produced, only copies are sold.[14] It is clear that the digitisation of everything from films to music, pictures and lecture courses, has generated all kinds of problems for industries that make their profits from selling those things (e.g. the film and music industries). And the root of the problem indeed seems to be that, once digitised, films, music, etc., can be copied and transported at almost no

---

[13] A rivalrous good is one that cannot be consumed by one person without making it impossible for another person to consume it. For example, my consumption of an apple makes it impossible for others to consume. Whereas the consumption of a non-rivalrous good cannot be easily prevented by many different people – a street light provides an obvious example of such non-rivalry.

[14] This simple situation of course become more complex if a distinction is made between knowledge, requiring a knower, and information, requiring only an informer, is made. Thus information can be stored in something that can be traded in the usual way without this essential property of sharing of knowledge being involved. But such arguments are incidental to the points I wish to make here.

cost at all, thus opening up the possibility of all kinds of ways to download a range of products for free.

Such arguments seem quite correct and are clearly identifying something significant about recent changes in economic activity, which in turn are responding to changes in technology. But it seems to be developments in the material realisations of code that is significant here. It is the changes in the material means by which software is processed and stored (processing chips and computer memory) and the means for connecting computers (fibre-optic cables, etc.,) that lie at the heart of the changes in economic activity noted above. Complex and well-documented calculating algorithms go back to at least the time of Euclid. But this seems of little use in explaining the relative explosion in mechanical devices, technological systems, etc., highlighted in Chapter 2 above, that have called forth the need for the modern meaning of the term 'technology' or, more recently, the digital revolution.

Similar arguments have been made recently within a variety of different literatures focusing upon digital technology. For example, prominent contributions to the philosophy of software, stress that code needs to be analysed 'as a medium that is materialised into particular code-based devices' (Berry, 2011, Manovich, 2002, 2013). This is also what seems to underlie the idea found for example in the dual-natures literature, that software is best understood as unfinished or incomplete technology (Kroes, 2012, p. 2). A similar idea is also to be found in criticisms of some of the posthuman literature. For example, Hayles highlights the question of how information is presumed to have, in her terms, 'lost its body' (Hayles, 1999). Hayles is primarily reacting to the idea she finds implicit in various posthuman contributions that it might actually be possible to download human consciousness into a computer. To do so would require human consciousness to be nothing more than patterns of information that can be materialised and dematerialised at will, or in any chosen location. Whether such materialisation takes places in a silicon body or a human body is of no consequence; information somehow coheres immaterially, ready for 'rehousing' in different ways. Hayes' desire to bring back the body into cultural discourse is pretty much the same as the desire of Berry and Manovich to bring back the material device into discussions of software and coding. Either way, disembodied or non-material phenomena seem relatively less important in explaining the digital revolution; it is the material dimension of computer technology that seems to explain its novelty.

I do not want to argue, however, that coding and software play no role in these new technologies. It does seem to me that computer technologies, based as they are on the processing and storage of code, do have various interesting properties. Alongside those features usually referred to, such as ease of copying, non-rivalry, etc., a feature I would highlight is its inherent differentiability or isolatability. There seems little in the world that is likely to be so precisely isolatable as series of 0s and 1s. Indeed, the procedure of digitising the world (transforming analogue information into bit strings) seems an excellent example of the what I term the 'moment of isolation'. The digitisation of a whole range of different features of the world are reduced to a series of 0s and 1s, which make all kinds of manipulation possible. The idea is wonderfully captured by Ellen Ullman when she talks of digitisation as the point at which 'human needs cross the line into code. They must pass through this semi permeable membrane where urgency, fear and hope are filtered out and only reason travels across' (Ullman, 1997). Not for the first time, reason is here associated with precise, logical statements that presuppose the differentiability or isolatability of the phenomena focused upon. This observation is also returned to in Chapter 8.

Similarly, code and software seem particularly conducive to the kinds of recombinability focused upon by those such as Ayres. Indeed, this is often thought to be an essential feature of coding, as Berry notes: 'coding is written through a process of collage, whereby different fragments of code (usually called "#includes") are glued together to form the final software product (this is the key principle behind software libraries and object oriented programming). This naturally undermines the notion of a single author of software, rather … programmers reuse old, reliable code wherever possible' (Berry, 2011, p. 40).

These themes are returned to in later chapters. For now, the important point I want to make is that software and code are not insignificant in the role that they play in modern societies. But they are able to play that role because of the material devices and artefacts in which they are always embedded. In this case, computer technology does not really provide a counter example to conceptions of technology, such as that of Ayres, in which material artefacts play such a pivotal role. Indeed, although computer technology has particular features that mark it off as special or unlike other technologies, in terms of the features I am highlighting as essential for technology, such as moments of isolation and recombination, computer technology, based on computer software and code, seems to fit very well within the account I am putting forward.

Concluding Remarks

To take stock, the previous chapter emphasised the importance of positioning for an account of the sociality of artefacts. This chapter began by raising the problem that if positioning determines what some artefact is, there seems no basis for deciding which are, and which are not, technological artefacts independently of how particular communities decide to position or use them. Here I am proposing that technological artefacts are different in that positional or deontic power is of relatively little use in understanding their use or functioning. At the same time, I am suggesting that this distinction between technological and other artefacts, as I have reformulated it, captures much that Ayres is concerned with when he distinguishes technology and institutions, especially in that icon-use tends to reinforce or reproduce institutional factors, tool-use tends to undermine them.

There are various reasons for my extended discussion of Ayres's work. In particular I believe that Ayres is at his strongest when he weaves this distinction into a general account of technological change over time. Here, as we see in Chapter 10, Ayres's comments about exponential growth, the principle of combination, simultaneous invention, etc., can in large part be seen as developing the implications of recognising both the importance of the isolative moment in technical activity and the relative isolatability of the intrinsic causal properties of material artefacts themselves, or put another way, the relative absence of the need for collective agreement, social relations, etc., for capable use of technological artefacts.

The question of why it is that changing technologies tend to disrupt and be inhibited by existing social relationships and structures is one that recurs throughout the technology literature in a range of different guises. Of course, for conservatives such as Heidegger or neo-communitarians such as Borgmann, it is the disruptive properties of technology that are 'bad' (undermining pre-industrial communities or simply disrupting the processes whereby communities invest their customs and practices with the knowledge they have gained), rather than good (as they are for Ayres because they disrupt the existing vested interests and biased hierarchies). But rarely is it asked, what is it about technology that makes it so disruptive? At least the beginnings of an answer can be found in Ayres's work, with his focus upon the relative isolatability and travel of material artefacts and the importance of, and growing opportunities for, recombination.

However, whereas Ayres suggests ontological differences in the constitution of different kinds of artefacts, he says little about exactly how they are used, or in what ways technological artefact use differs from other forms of artefact use. Similarly, there are many other kinds of artefact use that seem to appear to share the characteristics I have so far singled out. For whilst it now seems possible to say that certain artefacts such as passports, identity cards, and money should not be included within an account of technology, there are still many other objects that seem to be indistinguishable, such as food, art or toys. In order to make further distinctions, I argue, we need to return to the idea, suggested above, that technology is primarily used to extend human capabilities.

7 | Technology and the Extension of Human Capabilities

The main argument of this chapter is that another distinguishing feature of technology lies in the role that technological artefacts play in extending human capabilities. However, the sense in which I understand technology to *extend* human capabilities requires a fair amount of clarification. In order to provide this, I need to distinguish the argument I am making from similar ideas that exist in the philosophy of technology literature. The literature I have in mind here is that in which technology is conceived of as the more or less direct extension of human faculties. Specifically, I first discuss some prominent extension theories, including the classic accounts of Ernst Kapp and Marshall McLuhan as well as some more recent contributions. The intention is not only to clarify the sense of extension I have in mind but also to introduce some ideas that I return to later on. In the second section, I try to integrate these extension ideas into the general conception of technology developed in the previous two chapters before finally drawing out some implications of the account I defend. To repeat, the main task is to indicate how the idea of extension acts to mark off, at least partially, technological from other artefacts, along with suggesting some of the benefits that follow from such a demarcation.

Extension Theories of Technology

By extension theory, I mean any theory in which technological artefacts are conceived of as some kind of extension of the human organism by way of replicating, amplifying, or supplementing bodily or mental faculties or capabilities. This basic idea of extension recurs throughout the study of technology and is found in discussions of technology that go back at least as far back as Aristotle. The more systematic treatments tend to emphasise one or more of three features: a focus upon the direct,

This chapter is largely a development of Lawson (2010).

often very mechanical, extension of 'physical' faculties; a focus upon the extension of cognitive (especially information processing) capabilities; the extension of human agents' 'will' or intentions. I shall illustrate each feature by briefly referring in turn to the work of Ernst Kapp, Marshall McLuhan and David Rothberg.[1]

The first occurrence of a detailed and sustained example of an extension theory is that provided by Ernst Kapp (1877). For Kapp, technological artefacts are quite simply projections of human organs:

> the intrinsic relationship that arises between tools and organs . . . is that in the tool the human continually produces itself. Since the organ whose utility and power is to be increased is the controlling factor, the appropriate form of a tool can be derived only from that organ (Kapp, 1877, pp. 44–45).

All technology is, quite literally, a direct projection or 'morphological extension' of human organs. Throughout his book, Kapp is at pains to note how a wealth of different devices originate from such projections 'the bent finger becomes a hook, the hollow of the hand a bowl; in the sword spear, oar, shovel, rake, plow and spade one observes sundry positions of arm, hand and fingers' (ibid.).[2] The strength of Kapp's account is his tirelessly enthusiastic and detailed use of one example after another to support the claim that technological artefacts are little more that organ projections. Whole chapters of his book are given over to the more important developments of the time (for example, Chapter 7 is given over to the idea that the railroad is the externalisation of the circulatory system, and Chapter 8 to the telegraph, which is portrayed as an externalisation of the nervous system).

A variety of later works drew upon and developed Kapp's basic ideas. For example, Marshall McLuhan similarly conceives of technology as some form of extension, and shared many of the specific interests of Kapp, such as the railroad and telegraph, adding more recent interests in electronic media: 'during the mechanical ages we had extended our bodies in space. Today, after more than a century of electronic technology, we have extended our central nervous system itself in a global embrace, abolishing time and space as far as our planet is concerned' (McLuhan, 1964, p. 19).

---

[1] Examples of similar conceptions include 'organ-artefact circuits', (Feibleman, 1979), 'organ substitution' (Gehlen, 1980) and 'exosomatic organs' (Lotka, 1956).
[2] Cited in Mitcham (1994).

The crucial difference between Kapp and McLuhan is that the latter distinguishes two broad classes of extensions: of the body and of cognitive functions. Extensions of the body refer for the most part to those mechanical extensions that form the basis of Kapp's earlier work. McLuhan's emphasis is however slightly different, focusing more upon the isolation of particular properties and amplifying or augmenting these. 'What makes a mechanism is the separation and extension of separate parts of our body as hand, arm, foot, in pen, hammer, wheel. And the mechanization of a task is done by segmentation of each part of an action in a series of uniform, repeatable and moveable parts' (ibid.).

The senses, the central nervous system, and higher cognitive functions are not, however, defined as part of the body. It is in terms of these that McLuhan's analyses his central concern – the media – especially sight and sound. For example, the radio is long-distance ears. Electronic media are understood as extensions of the information processing functions of the central nervous system. Consequently, a human being in the electronic age is quite literally, for McLuhan, 'an organism that now wears its brain outside its skull and its nerves outside its hide' (ibid.).

This difference in emphasis leads McLuhan away from Kapp's insistence that the form of technological artefacts imitate the form of human organs. Instead, McLuhan argues that it is only functional properties of humans that are translated (in amplified form) to artefacts. Thus the focus is away from the role played by the projection of organs and functions onto artefacts. Instead, McLuhan's interest is in understanding the implications that follow for personal autonomy as bodily functions are taken over by machines, a point that is returned to below.

The work of David Rothenberg provides an example of the emphasis upon the extension of intensions. More specifically, technology is understood to be a process whereby intentions are realised via the extension of human 'aspects' that we understand the workings of:

A part of the human essence is evident in the things which we build, create, and design to make the Earth into *our* place. Techniques can extend all those human aspects for which we possess a mechanical understanding. Telescopes and microscopes can extend the acuity of our vision, because we know something about how our eyes perceive the world optically. But we cannot technically extend our sense of what is right, because we do not understand how this judgment operates (Rothenberg, 1993).

Rothenberg considers both thought and action as faculties that become extended in technological artefacts. The extension of action is, like McLuhan's bodily extensions, close to the work of Kapp and similarly Rothenberg's extensions of thought equate roughly to McLuhan's extensions of the senses, the nervous system and consciousness. These include artefacts that improve the senses (e.g. telescopes and radios); tools of abstraction that extend abstract thought and language functions (e.g. computers and calculators); and material extensions of memory (e.g. photographs and video). Rothenberg, like McLuhan, does not restrict his ideas of extension to projections of human organs. However, for Rothenberg, technological artefacts are not *primarily* extensions of human capabilities either. Rather, they extend human intentions. To talk of technology as 'an extension means that when we make something, we thrust our intentions upon the world' (ibid.). Intensions or desires are normally contained within our own organism, but as we create technologies, these technologies become carriers of our intentions, and hence extensions of them.

One of the few attempts to systematically compare and develop different extension theories is provided by Philip Brey (2000). Brey is concerned with the possibility of generalising the concept of extension developed by earlier contributors. Specifically, he is concerned that none provide a sufficiently restrictive sense of 'extension' according to which *all* technical artefacts can be claimed to be extensions of human faculties (that is, without counterexample). For example, whilst impressive, Kapp's accounts of artefact-organ pairings are ultimately unsatisfactory because many artefacts have no obvious origin in human organs: 'fishing nets, books, cigarette lighters, telephones, and airplanes, for example, do not have clear morphological similarities to human organs' (Brey, 2000, p. 66). Brey criticises others, such as Rothenberg for being ambiguous. Sometimes Rothenberg claims, along with McLuhan that artefacts functionally correspond to some human organ (which appears to be unsustainable) At other times it is only human intentions that are extended, in which case it is not a theory of the extension of human capabilities at all.

It is Rothenberg's ideas, however, that Brey most closely builds upon in his own formulation of an extension theory. For Brey, the emphasis is upon the way that technology extends the *means* by which human intentions are realised. Thus it is not human intentions that are being extended, but in trying to realise intentions human beings develop technological artefacts that extend the arsenal of means by which such intensions can be realised. Initially intentions are realised through what Brey terms 'the inventory of original means' – in order to change the world so that it conforms to our intentions, we have only our bodily and mental faculties available. Technical artefacts extend or add to these means.

Brey's account clearly highlights some of the ambiguities in the ideas of extension theories, and also undermines any claim to generality given the existence of important counterexamples. But this ambiguity is perhaps not avoided in Brey's own account. Specifically, it is not clear exactly what is being extended in Brey's conception. If it is the *means* that Brey has in mind, does it not make more sense to talk of simply *adding to* the arsenal of means? If extension is to connote the extension of the human agent, what exactly is it about the *agent* that is being extended? On Brey's conception, technology appears to be simply a distant 'means' to be utilised in some instrumental manner. In which case it is unclear how it is an extension in any obvious sense.

It is also unclear that such a sharp differentiation between intensions and means is especially helpful. If, rather, human agents are conceived of as ensembles of powers, or more specifically as centres of powers (Bhaskar, 1978), then intentions and means are simply part of the structural requirement for the possession of capabilities in any real sense. However, the broad idea of agents with powers possessed in virtue of the way they are structured, is hardly implausible. In this case, the question becomes: where does the internal structuring of the human agent end, given that many of a human being's capabilities are dependent upon the relations in which he or she stands to other people as well as to particular artefacts? Such attempts to demarcate need not detain us here. The point is that such boundaries are likely to be fluid and important for understanding what any human agent is able to do.

This idea of fluid human boundaries is of course a central proposition in the distributed cognition perspective (Hutchins, 1995). Here, although the focus is more narrowly on *cognitive* capabilities, the point is the same; the very capabilities that people have depend upon the relations in which people stand both to other people and to things. Of course, the manner of this dependence will change.

It may involve an important iteration changing the nature of the agent or simply be some kind of off-loading of capabilities (Salomon, 1993). Capabilities of the human agent become augmented and extended across time and space.

There is, however, a limitation to the above extension ideas which is worth focusing upon. This is that the ideas of extension tend to end in the acquired capability or the effects of the acquisition of that capability. But for the extension in capabilities to be realised, I want to argue, the artefacts or devices that are used to extend the capability must be positioned within both technical and social networks of interdependencies. To pursue this idea it is helpful to turn to a set of ideas that are not normally considered in this context, namely, those of actor network theory, and argue that it too is a form of extension theory, one with some important advantages.

Extension as Positioning

Perhaps the central proposition of actor network theory is that technological artefacts cannot be understood in isolation. Rather, technological artefacts take on their properties, characteristics, powers or whatever only in relation to the networks of relations in which they stand. This idea is often presented by arguing that artefacts *are* important social ties. For example, Latour even likens technological artefacts to the mass that physicists cannot find in the universe, which it needs to hold together (1994). In sociology, Latour argues, there has been a similar problem, namely, of finding the social ties that constitute or reproduce human societies. These missing ties, Latour argues, are technical artefacts, and they can only be understood in terms of the linking or relational properties they have, or the actual position they occupy.

Perhaps Latour's best known account of such linkages are to be found in series of 'mundane' examples (Latour, 1994). At one point Latour talks of his frustration at trying to persuade his son (being too old for a child seat but too young for a seat belt) not to sit in the, more dangerous, middle of the back seat – dangerous, that is, if braking were to occur too quickly. After failing with commands and the use of his right arm to restrain the boy, Latour finally purchases a padded bar to hold his son in place – thus the work done previously by his voice and arm are 'delegated' to a technical object (the padded bar). Such stories, and Latour gives many, are examples of a form of extension, in this case of the extension of the power of a parent-driver over his son.

Elsewhere, Latour talks of a mechanical door-closer (or groom). Here, sparked off by a sign pleading that, in the absence of a working groom could people please shut the door, Latour gives an account of the networks of use in which the door closer exists and operates. The account is wide ranging, moving from a discussion of what it would be like not to have gaps in walls at all, to the benefits of doors and hinges, through to the multiple kinds of problems that arise in attempting to discipline the users of the door to keep it shut after they have walked through it (thus ensuring a good warm temperature inside the building). Multiple 'delegations' are considered, in which small children could stand by the door to make sure it is shut (thing to human), through to the deskilling and displacement issues of introducing a mechanism which despite its imperfections (overly tight spring, hard-to-push hydraulics, etc.,) in some sense gets the work done (human back to thing).

Throughout, Latour is drawing out implications or lessons from his examples. Centrally, they are intended to show how morality or ethics are imposed on the user or 'prescribed' through use. Foucault's influence is of course noticeable here. But the process of enrolling things into networks is not always something to do with 'disciplining'. Rather, I want to suggest that these examples show that delegations of this kind tend to involve extensions of capabilities. But, more importantly, they involve extensions of capabilities by the positioning of artefacts in particular systems; these systems consisting in relations of interdependencies with their own built-in politics, asymmetries, etc. Thus, such examples illustrate the use made of the capacities and properties of material artefacts to extend the properties or powers of people (to discipline the son or to enforce door-closing, etc.,) through positioning.

It is difficult for Latour to talk of the properties of things in this way, however, because of his commitment to the 'flat happenings' or flat ontology that he defends.[3] But for the ontology set out above, such an account is a natural line to pursue. To recap, it was argued there that an important feature of the positions occupied by people is that rights and obligations attached to such positions enable all kinds of positional

---

[3] For more on why Latour finds it difficult to develop an ontology of things with powers and potentials see Harman (2009).

power. Rephrasing this somewhat, there is a positional extension of the institutional capabilities of the occupants. And much of this is no doubt intended. Thus, if community-sanctioned power (over others) takes the form of positional rights (and duties), then much of the intentional pursuit of power is likely to take the form of either attempting to occupy powerful positions, modifying the rights and duties or one's own position or seeking to create and occupy a new position with desired rights and duties (T. Lawson, 2012, p. 369). A complementary way to view the positioning of technological artefacts is, then, that they are used to extend human capabilities. But the extension here is one of extension through a mixture of harnessing the capabilities of things and positioning artefacts in particular systems.

It is this conception of extension, as positioning within existing systems, that I wish to utilise and develop. It both draws on the intuitions that we use technological artefacts to extend ourselves in some way, but combines this with the insight that such extensions are always dependent on the context of interdependencies that artefacts must be inserted into in order to function.

Relations Between Artefacts and Users

A variety of criticisms of extension theory exist that focus upon the relationship between technological artefacts and their users. An illustrative example is provided by Kiran and Verbeek (2010). Their main criticism is that extension theories presuppose an *external* relationship between the two. By this they mean that humans and objects are 'distinct ontological categories' that may interact but not be intrinsically connected (Kiran and Verbeek, 2010, p. 411). In other words, they suggest that relations between the two, in extension theories, are not in any way constitutive. This problem manifests itself in supposing some form of instrumentalist view of technology, a 'transparent' conception of human-world relations and a failure to recognise the role of technology in the constitution of human subjectivity.[4] I briefly consider each specific manifestation in turn.

---

[4] Similar arguments are made by Böhme, who formulates his criticisms of extension theories as being unnecessarily utilitarian (Böhme, 2012). Technology, Böhme suggests does not make action more efficient, it transforms it. In response, he adopts Foucault's concept of a *dispositif* in order to emphasise the idea that the

The instrumentalist view, which is focused upon in more detail in Chapter 11, is one in which technical artefacts are considered to be neutral. Human beings accomplish deliberate and premeditated plans *through* technologies. Humans and technical artefacts enter technical activities as predefined, the latter simply augmenting the powers of the former. One problem with this view is that technology so clearly shapes action. The question is whether this is because particular ideas or values become built into the design of much of technology (Winner, 1978, Feenberg, 2000), are enscripted to be used in particular ways (Akrich, 1992) or simply are such that they *afford* handling in particular ways (Gibson, 1977).

Extension theory, for Kiran and Verbeek, also supposes a transparent relation between humans and the world. Drawing upon Heidegger, they argue that using particular tools or techniques 'discloses' the world in a particular way; some things are accentuated and stand out, others slide into the background. The world appears to us in certain ways that are shaped by the tools we use. In this case our goals are also transformed by the tools we use as they present us with possibilities or opportunities that would not otherwise come about. In so doing, according to Kiran and Verbeek, tool-use changes the very nature of our subjectivity.

Whatever the merits of Kiran and Verbeek's criticisms understood in terms of the older more direct extension ideas,[5] they seem less appropriate to the kind of extension by enrolment that I am suggesting, and it is important to see why this is. Given the varied and complex mechanisms by which artefacts become positioned, there is nothing to suggest a very instrumental or transparent 'bolting-on' of artefacts to people as is being suggested. Indeed, this would appear to undermine the understanding of the social world outlined in Chapter 3. Moreover, the positional conception of artefacts put forward in Chapter 5, set within the transformational models of social and technical activity, seems particularly well suited to developing the idea that technologies provide the conditions of action, changing what is possible, expected and desired as Kiran and Verbeek suggest. Indeed, this is a crucial aspect

conditioning factor (in this case technology) makes some things possible, but also limits and shapes what is possible.

[5] Although it is debatable if there is the kind of heterogeneous conception of extension that is implied by Kiran and Verbeek's account, this is not the place to consider a detailed rejoinder to these arguments (although see Heersmink, 2012).

of the account I am defending. Not only is the positionality of technological artefacts reproduced and transformed through use, along with
other features of the core-social, but technological artefacts form the
material conditions and consequences of such action, where material
conditions serve not only to enable and constrain, but also colour and
constitute the action and actors involved.[6]

Kiran and Verbeek may still argue that the idea of extension forms a
relatively impoverished model for human-world relations. In response
to this, I can only say that the idea of extension is not intended to be an
exhaustive model of any kind. Rather, on the account being suggested
here, it is a component in a much more complex conception of technology, but only one component. This point echoes more recent criticisms
of my position. For example, Steinert, in providing a general 'stock
take' of extension theories, and for the most part suggesting a positive
inclination towards the formulation of extension theory I am arguing
for, is critical of the fact that this idea of extension, by itself, does not
'pick out technology specifically' (Steinert, 2015). This is quite correct,
but it should be clear also that this is not my intention. The idea of
extension is only a component of an account of technology, although as
I argue later, an important one.

As a component, however, it has several advantages that Kiran and
Verbeek, and Steinert, might seem well disposed to. For example, it is
likely to sensitise the enquirer to a disruption of interdependencies.
Change driven by technological artefacts is likely to be more disruptive
than other kinds of change. A crucial point here is that new technology
will not typically be equally accessible to all. Thus as technological
artefacts are introduced, the capabilities of some, but not others, are
extended. As such, existing networks of interdependencies are inter-
rupted, empowering some and disempowering others. Thus whilst the
notion of extension adopted here is one of extension through a mixture
of harnessing the capabilities of things and as positioning of artefacts in
networks of interdependencies, it is worth noting that there is an
inherently more disruptive dimension to the enrolling of technological
artefacts. As argued in the previous chapter, whilst the use of such
artefacts as passports or identity cards involves harnessing community-
based deontic powers, such use tends to reproduce, compound and

---

[6] As such, this notion of extension should also be acceptable to those such as
Böhme (see footnote 5).

reinforce existing relationships through use. This will tend to be the case for any activity which, in the manner detailed in the transformational models above, reproduces or transforms the position and status of particular individuals. However, the use of technological artefacts, which extends the capabilities of some with respect to others, has a much greater tendency to disrupt and undermine existing relations and networks of interdependencies. Without a clear distinction between the different roles or characteristics played by different activities and different artefacts, it is difficult to see how it would be possible to build up such a complex explanatory account in practice.

Concluding Remarks

The previous chapter focused upon the important role played by harnessing the causal capacities or features of technological artefacts. So doing, I argued, makes it possible to ground our intuitions that certain kinds of things, such as passports, money, identity cards, etc., are best not viewed as technological artefacts. But we still were not able to distinguish technological artefacts from such objects as food and toys. The category that usually rises to the fore at this point is use: technological artefacts are used *for* something; they are a means to a particular end or have some kind of function. But still this does not quite make the distinction. Political drama often serves the purpose of sensitising the population to all manner of 'issues', toys serve the purpose of play, food serves the purpose of sustenance, etc.

Rather, I want to argue that the idea of extending human capabilities, as developed in this chapter, is able to provide this last step in the definition. Specifically, I suggest a conception of technical activity as that activity which harnesses the causal capacities and powers of material artefacts in order to extend human capabilities. Eating food is not, then, a technological activity as it is intended for sustenance, survival or even enjoyment. Any item of food is a one-off act of consumption that has an effect; but there is no sense in which it is an extension via some kind of positioning in existing networks of interdependencies.[7]

---

[7] Although of course, there will be other situations where something like food is consumed for other reasons, such as a performance-enhancing drug. But the complication arises, and the possible understanding of the process as technological will hinge at least in part on whether the food is consumed in order to extend human capabilities.

Similarly play, on this account is not a form of technical activity. Clearly children use toys to *develop* their capabilities; hand to eye coordination may improve with play along with and understanding of how objects function, break, etc. But the point is that such toys might be taken away and the capability remains. This typically will not be the case with technological artefacts, and was certainly not their intended use in the first place. Capabilities are extended only as long as artefacts are positioned in particular systems. Moreover, much art is primarily concerned with harnessing the intrinsic properties of material artefacts, but even when it is, it is not engaged in for the purpose of extending capabilities in this enrolment sense. It may be simply for some (aesthetic) consumption, play, fun or whatever, but once it becomes simply for the sake of instrumentally extending our capabilities it is not clear that it is art in any sense. In short, I am suggesting that when we use technology, we harness the capabilities of material artefacts in order to extend our capabilities.

I want also to indicate the usefulness of this account by suggesting, somewhat more speculatively, certain lines of inquiry that are opened up by a focus on extending human capabilities. To this end, I wish to return briefly to some of the central themes of the extension theories referred to above and in particular to some of the questions and considerations that arise from an extensions perspective, amending these in the light of the positional conception of extension I have argued for.

One ever-present question in discussions of technology, from Heidegger to the Amish, is how are we ourselves transformed by using some particular technology; what does using some particular technology make us become? Despite the reservations of Kiran and Verbeek, this issue arises in the extension literature in a variety of different forms. For example, Rothenberg and McLuhan ask what we are extending ourselves for and which artefacts extend us in ways that are desirable or compatible with that which we most desire (or with the kinds of us that we wish to be)? More generally perhaps, given the particular focus suggested here, we can ask what kinds of things do we wish to be capable of and, of course, are we happy with others being capable of? But, perhaps more importantly, the point is to distinguish a general need for increasing capabilities or realising potential capabilities and the effects that such capabilities might have.

In this light, if we take Rothenberg's ideas seriously, a related issue is whether we recreate ourselves, if only partially, in our technology?

If technology provides a mirror to view ourselves, then does our conception become one-sided or mechanical, implicitly devaluing all of the human faculties that cannot be so extended in technological artefacts? However, it is not simply that we extend those aspects of ourselves that we understand, as Rosenberg and others claim. Rather, extension on the account provided here involves moments of isolation and reconnection. Thus our extensions are concerned with functional properties that may indeed be isolatable at least momentarily and as argued in the following chapter, there is a real possibility that much of what we have been terming the sociality of artefacts may be 'filtered out' in this process. The result is for a tendency for the *lifeworld*, if only influenced by matters of technical expediency or efficiency, to become mechanistic and a source of an impoverished conception of ourselves. However, an immediate corrective can be found in the role of positioning on this account. To make technologies work they must be positioned, which always involves some kind of enrolment or reconnection to existing networks of interdependencies. Perhaps the implication of this is that there are likely to be different waves of differing conceptions of ourselves. But a more likely implication, given the institutional separation of these moments in contemporary capitalism, is the emergence of groups of technicians or engineers most concerned with the isolative moment, where mechanistic conceptions of humanity will tend to dominate. Such groups will evolve different cultural conceptions, values, etc., that may form and develop relatively independently of the conceptions of those concerned with reconnecting, embedding and enrolling artefacts into our lifeworlds. Thus we have further reasons, to those advocated by antitechnocracy theorists (e.g. Postman, 1993), to be wary of deferring decisions about which technologies should be pursued to communities of such experts. This theme is taken up in Chapter 9.

Another related idea is that the extension of capabilities also involves some degree of off-loading of certain functions, tasks or abilities to machines. The idea that such a process serves to impoverish us in some way is a familiar one within the work of philosophers of technology such as Borgmann (1984), Illych (1973) and Marcuse (1964). Each author has argued that this process is one by which contemporary technology reduces human beings to a state of greater passivity. Translated to an extension perspective, the worry (as articulated in McLuhan's work) is a lack of personal autonomy and concern about how the intentions of different members of society are likely to be represented. The extension-enrolment perspective suggested here, would amend the general thrust of this concern in that this process be understood at least in part by the way that extending our capabilities commits us to, or encourages us to invest in, particular networks of interdependencies. Whilst absolute isolation or passivity will not generally be the results of such enrolments, such features as the distancing of ourselves from the effects of our actions and the disruption of existing networks of interdependencies will involve moments of disconnection and a general tendency towards it under certain circumstances. For example, if new technological artefacts emerge at a rate that is too fast for users to be able to position or enrol such technologies so as to result in very healthy or meaningful 'reconnections' (connections that really do incorporate what we learn about the 'good life' on a daily basis), then a degree of dislocation or disconnection would seem to follow along with the experience of 'out-of-control-ness' that has been the focus of what so many authors often termed technological determinism (Lawson 2007). This idea is returned to in Chapter 10.

One last idea, implicit in much of the above is that in extending our capabilities, technological artefacts also distance us from the results of our actions – from emailing or texting rather than conversing with someone in person, to using a spanner rather than our fingers to undo a frozen nut, technologies insulate us to some extent. But, given the enrolment perspective adopted here, the same act tends to connect us too (literally, to communication networks, or positionally, to communities of tool users). The relation between user and artefact is not one-way, as critics of extension theories often seem to suggest.

Lastly, we are now in a position to return to an issue raised at the end of the previous chapter. Technology has always been considered to play a central role in prompting or driving huge social changes, even though the character of technology's influence has varied from author to author. But, as noted, rarely is it explained why technology might have such a disruptive influence or investigated whether there are commonalities about the processes by which such disruption takes place. Some of the answer to this question was provided in the previous chapter. But another element follows from an extension perspective which emphasises that existing networks of interdependencies are disrupted as some capabilities are extended. The introduction of some technology or other involves the extension of some capabilities, empowering some whilst making others disempowered or even redundant. Thus a central task will be to question whose capabilities (to control, to defend, to manipulate, etc.,) are being extended, and any implications that follow. Technological advance is often portrayed as inevitable, unambiguously progressive and potentially good for all. But this emphasis on extension as enrolment, and so on the disruptive extensions of the power of some, is a useful corrective to such thinking and at least part of the story to which various social critics of technology, such as Marx, have attempted to draw attention. Technical change is far from neutral and deciding whether or not to embrace or welcome a particular technology should not simply reduce to a decision about how efficiently some technology performs a particular task. An extension conception provides a fresh perspective on such issues as well as helping to delimit a conception of technology in the first place.

8 | Technology and Instrumentalisation

The last few chapters have focused on different aspects of an ontology of technology. The aim of this chapter is to bring these different strands together and to indicate some of the advantages of the resulting conception of technology.

In Chapter 2 the argument was made for a multipart definition of technology, which covered not only the more traditional, etymologically more consistent idea of technology as a study, but also accommodated the changes to the term's usage that led to it becoming a keyword in social theory. In short, technology can refer to a study, or the ideational or material results of such study, with popularity in the term rising as the latter component came to be emphasised. To employ such a multipart definition, however, may serve to detract attention from the real 'action' that gives rise to technological artefacts. Rather than the study of or results of study, the main point of interest is the action of making or facilitating the coming into being of technological artefacts and their use. Thus, an ontology of technology needs to be focused squarely upon those activities that give rise to technological artefacts. Accordingly, in the account above, a central feature identified is the process whereby causal factors can be isolated and recombined in order to produce material artefacts to be used for practical purposes, where being used for practical purposes here involves harnessing the capabilities or capacities of such artefacts in order to extend human capabilities.

In order to illustrate these ideas, and suggest that they accommodate many important contributions from the philosophy of technology, I shall focus in this chapter upon the work of Andrew Feenberg and in particular on his ideas about instrumentalisation. Adopting such a focus here is helpful for at least two reasons. First, Feenberg's work very usefully situates (and in places helps to make sense of) some of the ideas from the philosophy of technology in which I am most interested. Secondly, given how similar some of Feenberg's ideas turn out to be to those developed above, underlining the differences between Feenberg's position and that adopted here not only serves to clarify the argument being made but indicates the kinds of directions into which it might usefully be extended.

Instrumentalisation Theory

Feenberg's conception of technology is best understood as the latest in a tradition of critical theory stemming from Marx to Lukács and Marcuse, but which accommodates phenomenological, especially Heideggerian, contributions along the way.[1] Feenberg formulates his conception of technology in terms of what he calls primary and secondary instrumentalisation (Feenberg, 2000, 2002, 2010). Put very simply, and in such a way as to highlight similarities with the above, this distinction is between two different levels or moments of the process in which technologies come into being. Primary instrumentalisation is intended to capture the separating off and refinement of affordances of technological artefacts, whereas secondary instrumentalisation is concerned with the realisation of such objects as working devices in actual networks and other environments.

More specifically, primary instrumentalisation is understood to involves the 'de-worlding' of objects in order to reveal particular affordances (Feenberg, 2010, p. 72). Feenberg uses the term de-worlding to combine two ideas. On the one hand, it draws attention to the fact that things or aspects of things are artificially separated from the contexts in which they were originally found. But at the same time, de-worlding invokes the Heideggerian idea that this separation also involves an instrumental or utilitarian evaluation or 'revealing' of the aspect or object in question: 'nature is fragmented into bits and pieces that appear as technically useful after being abstracted from specific contexts' (Feenberg, 2000, p. 203). Coupled to this is the idea of 'reduction' in which the de-worlded objects become simplified and so reduced to those properties and forms that make them most useful. For example, when a tree is cut down, de-worlding refers to the separation of the thing from its context in which it provided shade, a habitat, etc.

---

[1] Feenberg's relationship to critical theory is explored in Lawson (2017). For now, the focus is more narrowly upon his conception of technology, especially as elaborated in his account of instrumentalisation.

But then, the tree is reduced to useable form, in that branches and bark are removed and the tree is cut into planks of a particular size for use in some kind of construction.

Primary instrumentalisation, Feenberg argues, can also be understood in terms of its effects upon the human subject. The term 'autonomisation' is used to capture the idea that users of technology are relatively unaffected by the objects they use, by distancing or deferring feedback from that object. For example, when a gun is fired the small 'kickback' experienced by the person pulling the trigger is nothing compared to the effects of the bullet fired. Primary instrumentalisation also involves some idea of positioning, although a rather different one to that used above. Feenberg intends the term positioning, or positioning with respect to something, to capture the idea that we cannot transform everything. In his usage, the term is used to capture an aspect of what I have described above as harnessing. For example, gravity and the boiling point of water are phenomena that we cannot control but that we must navigate or position ourselves in relation to.

At the secondary level of instrumentalisation, technological artefacts are integrated and combined with each other and other factors, so as to produce a world of working devices and systems. Feenberg often uses the term *enrolment* to capture the idea that such integration requires a range of different kinds of technical and social networks to be accommodated.[2] More specifically, Feenberg again distinguishes four aspects of instrumentalisation, these being termed systematisation, mediation, vocation and initiative. Systematisation involves making a series of connections that result in the device being made to 'work', such as adding different devices together and inserting these within working technical networks. The idea of mediation in this context refers to the ethical and aesthetic dimensions of such recombination where objects are invested with the kinds of 'secondary' attributes lost at the 'de-worlding' stage. For example, the motor car is designed to slot into a system of private ownership, be driven on particular kinds of roads and use particular kinds of fuels (systematisation), but also to appeal to our tastes, in the form of shape and colour of the car itself, and values, for example relative environmental impact (mediation).

---

[2] Although Feenberg credits the term to actor network theory, it is not clear that the rather vague way the term is used by actor network theorists actually contributes anything to Feenberg's account.

The term vocation is used to capture the sense in which involvement with technological artefacts does act directly upon the agent, reforming and constructing the identity of the agent. Obvious examples are premodern craft activities, which provide forms of community and pride of belonging, rather than simply prescribing a form of interaction with some particular objects. Lastly, the term 'initiative' is used to capture the idea that human agents are never completely determined by their interactions with devices – there is always room for the autonomy of some kind (Feenberg, 2010).

Feenberg emphasises that the connection between these moments is a dialectical one. Each of the four aspects of primary instrumentalisation can be understood to be compensated for in some way by one of the four aspects of secondary instrumentalisation: systematisation acts to recontextualise, establishing new connections; mediation serves to attach aesthetics and meanings lost at the reductive stage; vocation reconnects the user of technology with the effects of its use; initiative closes the distance between the user and technology opened by the need for positioning. This idea of compensation and reintegration is important for Feenberg as it is the primary means by which some normativity enters into his account. Under capitalism, these forms of compensation are largely constrained, and a better society would be one in which these forms of compensation are freed up. At present, Feenberg argues, secondary instrumentalisations are largely afterthoughts. A better society would make them central (Feenberg, 2010).

Instrumentalisation and Ontology

Feenberg's account is generally well received. It is persuasive, not only because it grounds many of our intuitions about technology, but because it accommodates and combines a range of insights from theorists of technology from Heidegger and Habermas to Latour and Foucault in a sympathetic and constructive manner. Feenberg's account is not, however, explicitly ontological. For the most part this reflects Feenberg's own roots in the phenomenological tradition, which quite consciously has sought an alternative to any kind of ontological orientation.[3]

---

[3] Having said this, it is fair to say that Feenberg's successive presentations and developments of instrumentalisation do seem to be moving in the direction of more explicit ontology, for example see Feenberg (2014).

However, it should be clear that instrumentalisation theory is very close to the account given above on a number of points, even if it is derived from different sources. By far the most obvious similarity is the role in both conceptions of some idea of isolation and the distinction made between a moment of isolation (de-worlding) of causal powers and one of their recombination (reintegration). There is also sensitivity to the myriad ways in which technological artefacts are positioned, often in complex and diverse ways, in order to work. Furthermore, there is an explicit attempt in both accounts to combine the sociality of technology with its technicality, its need to function in technical ways, and the relation between the two. It is helpful to consider each of these similarities briefly in turn.

Feenberg stresses, especially in his most recent versions of instrumentalisation theory, that these levels or moments are intended as *analytical* distinctions. It is not, for example, that cutting down the tree is simply a matter of primary instrumentalisation and using it to build something is an instance of the secondary instrumentalisation. The cutting down of a tree does indeed separate it from its context, but secondary considerations are in play right from the start. The cutting of the tree conforms to various standards or specifications concerning the size and shape of the planks cut. Technical, legal and aesthetic considerations determine which kinds of trees are to be used.

It is also clear, however, that primary instrumentalisation is about actually isolating things. Although Feenberg presents his ideas about primary and secondary instrumentalisation in different ways, some conception of isolation is always central. For example, when he talks of primary instrumentalisation initiating the process of world-making, this is 'by de-worlding its objects in order to reveal affordances. It tears them out of their original contexts and exposes them to analysis and manipulation' (2010, p. 72). The similarity to the role played by isolation in controlled experiments should be clear. By isolating the causal properties from the empirical flux in which they are always manifest, their operation is revealed. And of course, as in controlled experiments, success depends upon some things actually being isolatable. As noted, this is one point at which Feenberg's reading of Heidegger is most apparent. Objects that are recombined to form a working device or system already take the form that they do because they are the results of previous shaping and moulding to fit with some functional conception of what they might be used for. In Feenberg's terms, they are formed by the functionalisation they have been exposed or submitted to (Feenberg, 2010, p. 76). This idea is perhaps most pronounced when Feenberg is commenting upon Marcuse's distinctions between holistic and mechanistic views of human and natural systems. Here, although formal abstraction appears to be neutral, this overlooks the difference between the 'extrinsic values of instrumental subject and the intrinsic telos of an independent, self-developing object'.[4] Put more prosaically, some things are isolatable but other things are not.

Feenberg states clearly in several places that not everything survives the kind of transformation that primary instrumentalisation involves. Drawing from Heidegger once more, the idea is that much that is crucial to our being is in fact lost in this process of reducing everything to standing reserves or resources ready to be used in some or other technical device. It is of course this aspect of technology that Heidegger finds most impoverishing and accounts for his generally dystopian vision of a modernity in which technology dominates.[5]

Feenberg and Heidegger's arguments here require the idea that some but not all features of reality are actually *isolatable*. The emphasis upon de-worlding gives this moment a very phenomenological inflection. But some things are quite literally removed from their contexts, whilst for other things this is simply not possible. I argued in Chapters 3 and 4 how it is difficult to make sense of much that we know about the natural sciences, and in particular the role that replicable experiments play in such science without supposing that many of the important explanatory mechanisms that it deals with are actually isolatable. However, the focus upon social ontology suggests that much in the social world is not isolatable, which explains why the status enjoyed by controlled experiments in some sciences (such as physics) is very different to that of other sciences (such as economics). If the possibilities for isolation are so variable, we have an ontological grounding for at least some of those aspects that cannot survive the primary instrumentalisation process. Indeed, if much of the social world is not isolatable in the same way, then much of the sociality that surrounds

---

[4] Feenberg's use of the term 'formal abstraction' is close to the method of isolation as discussed in Chapter 3, and so on my terminology is an alternative to abstraction.

[5] This is, of course, very similar to Ullman's observation, discussed above, that when crossing the line in to code 'urgency, fear and hope are filtered out and only reason travels across' (Ullman, 1997).

and embeds a particular object or mechanism, will, in Feenberg and Heideggers's sense, not survive this form of instrumentalisation.

The idea of isolatability is clearly of most concern to the primary moment of instrumentalisation, but there are implications also for secondary instrumentalisation. This latter moment is one of reconnection of different components, devices and mechanisms within technical and social systems. But the isolatability of things makes a difference to how such reconnection can proceed. Although secondary instrumentalisation can take on different forms (such as systematisation and mediation), a crucial factor underlying the kinds of instrumentalisation that are possible is the isolatability of the things being combined. In this case, we can distinguish, at one end of the spectrum, relatively atomistic recombination, which amounts to little more than a mechanical aggregating or reordering of distinct devices, functional properties, etc., from, at the other end of the spectrum, organic recombination in which devices are inserted within systems or wholes (thus taking on different meanings, relations, ethics, aesthetics, etc.,) that cannot be treated or understood in the same way. As noted above, it is the more atomistic recombination that features centrally in Ayres's conception of recombination and underlies his explanations of technological development in terms of his principle of combination. Similarly, it is these features that underlie McLuhan's ideas about segmentation of tasks into a 'series of uniform, repeatable and movable parts' (McLuhan, 1964).

The second source of similarity between instrumentalisation and the account highlighted above is the importance attributed to different kinds of positioning or enrolment. An interesting example here is the way that Feenberg uses distinctions between the different aspects of each moment to account for widely held observations about the rough features, or changing outlines of modern societies. Thus it may be possible to understand differences between modern and premodern societies in terms of the changing relative importance of systematisation and mediation. For example, Feenberg suggests that in premodern societies, technical networks are relatively loose and their reach relatively small. Such networks crucially depend on deep ethical and moral enrolment, whereas modern societies can be characterised by relatively tightly connected and widely extending networks of interdependencies where enrolment is less entrenched.[6] In Feenberg's terms, the

---

[6] In advancing this argument Feenberg draws upon Latour (1993).

commonly made criticism of modern societies (that the extensive reach of such societies is achieved at the expense of a fragmentation of customs, communities, etc.,) is understood as at least in part to do with an undue weight given to systematisation above mediation.[7] From a more ontological perspective, such a transition might also be a possible mechanism whereby a more atomistic orientation within modern societies is replacing the more organic orientation of premodern societies, or a preoccupation with using methods of isolation rather than abstraction to understand the world. These ideas are returned to in later chapters.

Similar implications can be drawn out with respect to the characterisation of skills-based, tool-using technical activity in terms of isolation and recombination of technology. For example, Tim Ingold's well-known example of the weaver can be used to suggest a similar argument (Ingold 2000, Chapter 18).[8] For Ingold, the activity of the weaver demonstrates that making is not necessarily a simple process of human beings putting some explicit plan or design in to action. Rather, it involves a continual iteration between some conception of end-use and the materials engaged with feeding back on to the process of making. But this example could as easily be used to show that in certain contexts the necessary moments of isolation and reconnection often do take place together, even tacitly. By contrast, mass production is most often associated with an explicit and even institutionalised separation between processes of isolation and reconnection – where design or research departments tend to become quite disconnected from the details of how their (primarily isolative) research will be used by other designers (i.e. recombined with other technological artefacts into useful things), which are, in turn, disconnected in more far-reaching ways from those who may actually use the objects produced (contextualising or embedding these objects in particular social and technical networks).[9]

The last example of similarity between instrumentalisation and the account being put forward here is the insistence that technology is both social and non-social, as well as suggesting some kind of transformational relation between the two. Such distinctions are not always clear in Feenberg's work. For example in some places he simply asserts that

---

[7] The reasons why this may be the case are taken up in Chapter 8.
[8] See also (Lawson, 2008).
[9] This point is returned to in Chapter 10.

primary instrumentalisation relates to the non-social or 'functional' and the secondary relates to the social (for example, see Feenberg, 2010, p. 72). But this is not as simple as it seems. For the most part Feenberg is rather underlining that what he is terming the functional, that is the need for the device to actually work, is not something that can be reduced to the social. He is seeking to avoid a crude social constructivism in which the functional workings of technologies play little role. Rather, Feenberg emphasises the idea that in the first moment of instrumentalisation, different devices, or mechanisms are brought together in ways that reflect the values, desires and intentions of those designers and all the groups that have had some say in the nature of the design. Such values, intentions, etc., then become materialised in the very structure of technological artefacts and so can be understood to be exerting a continuing influence over technological activity both via the kinds of enablements and constraints noted above but also via what Feenberg terms the technical codes of operation built into and mediating their use.[10] In so doing, Feenberg's account becomes very similar to the transformational account provided in Chapter 5.

In order to make such points, Feenberg also draws upon the work of Simondon, and in a similar way to that above. In particular, Feenberg develops Simondon's conception of progress or technical advance that incorporates the idea that technologies become accommodated to their social and natural environment. Perhaps the main point of ontological significance here is that a range of social phenomena become materialised into technologies in different ways, but in so doing become different – they obtain a different mode of existence. One notable feature of Marx's contributions on technology, discussed in Chapter 11, is how not just values, intentions, etc., become materialised in this way, but social relations themselves.

To say that values, intentions and even social relations become materialised in this way is to talk of a process in which social things (such as values, relations, etc.,) take on a particular kind of material existence. As such, this process generates a form of endurability and travel to those otherwise precarious or fragile aspects of the social world that normally depend upon the continued activities of human

---

[10] Thus the use of technological artefacts is prescribed not only by the social relations implicated in an object's 'position', but by the rules of use literally built into the object itself. For an expansion of these ideas and their implications, see Feenberg's discussion of technical codes (e.g. see Feenberg 2002, pp. 20–21).

actors for their own existence. Thus, and this seems to be centrally important for an understanding of the nature of technology, technology is the site in which the social achieves a different mode of existence through its embodiment in material things. In other words, there is another sense in which the term 'transformational' is relevant to the use of technological artefacts. Such activity can be viewed as an important site in which the social is *transformed* into the material. This, again, seems to be an idea to be found within instrumentalisation theory.

Isolatability

If there are significant overlaps between the account of technology given in earlier chapters and that contained in Feenberg's instrumentalisation, there are also significant differences, many of which follow from the very different routes by which each has emerged.

Perhaps the most central point to make here is that there seem to be rather different ontological presuppositions about the social and non-social worlds. As noted, Feenberg's project has been far more explicitly phenomenological than ontological. But if we attempt to work backwards to the kind of ontological assumptions he is implicitly working with, it would seem that most of reality, following Marcuse, is understood holistically and there is little attempt made to distinguish situations where this may be more or less appropriate. This contrasts with the position I am defending here in which the crucial point is that reality is not always isolatable to the same degree, some sections or aspects of reality being more differentiable or isolatable than others.

Such differences have several implications. First, although instrumentalisation theory suggests that some things may not survive the process of primary instrumentalisation, there is little attempt made to develop this point or to explain why this may be so. Of course, on the account above, this 'filtering process' takes its character from the differences in isolatability to be found in reality. It is the processual, internally related character of much of social reality that makes the isolation of some things impossible in primary instrumentalisation. I suspect that Feenberg would hardly disagree with this explanation. But setting things up in this way serves to de-emphasise the similarities in the way that instrumentalisation theory might be extended to the social world. For example, in generalising instrumentalisation, Feenberg argues that capitalism is a system in which human beings are isolated and recombined in pretty much the same ways that material devices or components are. Whilst the exercise of extending instrumentalisation theory in this way is certainly very suggestive and there certainly seems scope for some overlap (another topic focused upon in Chapter 11), there are important limits to the way in which such isolation and positioning can be seen as equivalent in different domains. For example, human agents are not really isolated from their social context, their communities and backgrounds, at least not in the same way. Particular social systems may indeed attempt to separate and isolate particular social actors, but there are limits to such attempts which are not relevant if we talk about the isolation of causal factors in controlled experiments or in the development of new technological devices. The kinds of limits I have in mind here are well captured in discussions of the relative commodification of labour. For example Polanyi's use of the term 'fictitious commodity' to describe labour, seems to be making exactly this point (Polanyi, 1957).

Similarly, ideas of positioning (in Feenberg's sense) seem to be quite different for the social and non-social worlds. Positioning oneself with respect to gravity or the boiling point of water, seems quite different from positioning oneself with respect to the laws of markets or bureaucracies. Crucially, positioning oneself with respect to the operation of markets or other social forms, whether there are indeed laws or not in the same way, seems to depend on how particular agents are positioned (in the sense of the social position occupied). As a consumer with little money, or as the head of a hedge fund, different agents will of course be able to 'position' themselves with respect to the workings of different social institutions, such as markets, in different ways. Such differences will also be important for how human agents can extend their capabilities. As noted, both technology and social institutions can be used to extend human capabilities, but the manner in which such extension works and can be limited or amplified is also quite different as argued in the previous chapter.

Similarly, the inherent scope for isolatability has implications for the call for more integrative/compensatory relation to technology given the exponential proliferation of devices discussed in Chapter 6. If there is indeed such a proliferation, which follows from the increased scope for the recombination of devices, then it would be expected that there are essentially built-in limits to the possibilities for integration and compensation essential to secondary instrumentalisation, which are in addition to those political constraints that are the focus of Feenberg's work. This is another topic taken up in the following chapters, but put simply, integration and compensation take time and effort. With the proliferation of new devices there is less opportunity for the compensatory mechanisms to take place, thus adding constraints to secondary instrumentalisation, which are in principle quite separate from those that instrumentalisation theory identifies in the workings of capitalism.

Lastly, if reality is not equally isolatable or differentiable, then this opens up the possibility that individuals can have differing capabilities in terms of understanding or mediating differentially isolatable domains. In other words, some people may find situations where isolatability is possible, easier or more difficult to engage with or understand than other situations. These differences throw an important extra element into the dynamic of compensation that Feenberg focuses upon. Again, this is another idea that is returned to in the following chapters.

To repeat, these are all issues that following chapters attempt to address. For now, the point is simply to underline that they are issues that do not suggest themselves from within instrumentalisation theory, given the different orientation of that theory and the relative neglect of differences in isolatability.

Concluding Remarks

Let me take stock and summarise the broad features of the account of technology I am suggesting, before moving on to the applications of these ideas in the following chapters. The definition of technology I am employing has different components, as study and as the results of study, especially realised as material artefacts. But the main focus has been upon an ontological account of how these artefacts come about. A central part of this is the process whereby causal factors can be isolated and recombined in order to produce technologies, harnessing the intrinsic capabilities of material artefacts in order to extend human capabilities.

To understand how a working device, or system of devices, comes into being requires a complex ontological story. However, it is possible to conceptualise this story, much as Feenberg does, in terms of the articulation between two main moments. First, there is a moment in which affordances or causal powers of things are isolated and transformed. This is the moment which bears most relationship to

experimental science. But rather than isolate or retroduce to the existence of different causal powers, such powers are actually isolated,
separated in such a way that they can be recombined. Thus technology,
as study, clearly bears most resemblance to those aspects of science
most concerned with the creation or manipulation of (relatively) closed
systems, or controlled experimental conditions. The second moment
consists in the different ways in which such (re)combination can take
place. And indeed, a range of different kinds of reintegration are
required for some device to actually 'work'. At one end of the spectrum
we have fairly mechanistic or atomistic recombinations of isolatable
causal properties into working devices. This is Ayres's process of
recombination. But even here, at the most mechanistic level, recombination always involves the internalisation of functions and uses, more
or less as identified by Simondon. But for devices to work, however,
they must also be positioned. Reconnection to often extensive technical
networks is also required alongside a more affective *enrolment* within
our systems of relations and practices.

Furthermore, much of technology's peculiar character comes from
the fact that some things are more isolatable than others. Thus the basic
intuition that technology combines or exists at the interface between,
the social and non-social, is grounded. But most importantly, it is the
differential isolatability between social and non-social features of technology that I want to suggest accounts for much of the peculiar character and dynamic of technology.

If technological activity can be understood broadly as any engagement with technological artefacts, such as with design, refinement or
use, it is also possible to identify some differences in technological
activity in relation to other kinds of social activity. As noted, technological activity can of course be understood to have the same broad
features as any other kind of action, namely, it is undertaken by people
acting intentionally, in conditions not of their own choosing but
transforming the materials to hand and not reducible to or derivable
from the object of that activity, in this case technological artefacts (i.e.
technological subject and object are different kinds of things).
However, technological activity is primarily transformational in
a different sense to other activity focused upon in the transformational
model of social activity referred to above. Whilst positional characteristics of different technological artefacts are reproduced and transformed through human activity, this process is relatively less

important where technological artefacts are concerned. In using a hammer, I reproduce its status within a particular community as a certain kind of thing. However, such processes are far more important for objects such as passports or cash, which only have the causal properties they do because of their continued use within a particular community. Put another way, the most significant causal powers of technological artefacts are harnessed rather than reproduced or transformed in use.

In the act of constructing technical artefacts, however, technological activity does engage in a slightly different form of transformation where all kinds of social phenomena, such as values, ideas, rules, relationships, etc., are effectively solidified (internalised, encoded or inscribed) into the very structure of working devices and components. Thus the social is transformed into the material, and some things are given a different mode of existence, in ways which often go unnoticed.

This diachronic emphasis can be set against a more synchronic one. In particular, when technological artefacts are used it is primarily the causal capacities or powers of these artefacts that is harnessed in order to extend human capabilities – and the kinds of deontic powers central to social positioning are of little significance. This much is a crucial part of understanding how technical artefacts, and so technology, can be distinguished from other kinds of things. The focus upon the importance of the intrinsic capabilities of things, grounds the importance of a focus upon isolatable causal properties, mechanical recombination, transculturality, etc. The emphasis upon extension through positioning focuses attention upon interdependencies, collective action, social positioning, etc., in a way that foregrounds issues resulting from asymmetrical access to power and regular disruption to those patterns of interdependencies.

To repeat, at the heart of each of these components of technology, lies the relative isolatability of different causal mechanisms. Exactly how much isolatability there will be must ultimately be an empirical matter, there is probably not much more that can be said a priori. But such differences in isolatability, which lie at the heart of the different status of controlled experiments in natural and social science, also lie at the heart of many of the novel or surprising features of technology. Differential isolatability is as relevant for the diachronic as for the synchronic features of technology, albeit in slightly different ways. Over time it constrains the nature of the isolative moment in design and invention as well as the mechanistic features of recombination. But much is not isolatable, thus giving rise to the filtering-process aspect of technical design and invention and the ambiguity most of us seem to share with respect to the functionality of the resulting technological artefact. In other words, such artefacts result from processes of isolation and recombination, generating a ratchet-like tendency for ever more complex, integrated objects with ever more prescribed uses.

9 | *Technology and Autism*

It is uncontentious to suggest that those with autism tend to have some kind of special relationship with technology. An abundance of reports from parents and clinicians take it largely for granted that children with autism are attracted or drawn to various kinds of technological devices (Colby, 1973). More formal evidence for such a relationship has come from research into the use of assistive technologies (Wainer and Ingersoll, 2011, Goldsmith and LeBlanc, 2004). Elsewhere, support comes from research into the profiles of autistic children, where a disproportionate number of those with autism are male and offspring of engineers and mathematicians (Wheelwright and Baron-Cohen, 2001). Support is also found in claims that there are dramatic increases in the prevalence of autism in high-technology regions such as Silicon Valley (Silberman, 2000).[1]

A central feature of these reports is the idea that technology presents a more comfortable or more manageable interface between those with autism and aspects of a social world that are often experienced as mysterious and unsettling. Whilst a few studies exist that focus upon implications that follow from such a relationship for our understanding of autism, as far as I know there are none that ask if such a relationship might tell us something about the nature of technology. This is the general motivation of this chapter.

More specifically, this chapter summarises the main features of autism as it is currently understood along with a discussion of prominent theories used to explain it. An account of these broad features is then given in terms of the conception of technology developed in preceding chapters, focusing upon the nature of the relationship

---

[1] Silberman and many other commentators in fact talk of an autism epidemic. However, the use of such terms is increasingly contested by those arguing that it is unhelpful to portray autism in such terms. Indeed, Silberman seems to have more recently changed his mind about this portrayal of the situation (Silberman, 2015). Also see note 4.

between those with autism and technology. A central focus is upon the different kinds of isolatability that exist in reality and that are particularly relevant to an understanding of technology. My main argument is then that difficulties in coping with inherently unisolatable phenomena both prompt an attraction to technological devices and encourage particular kinds of relations to it. Finally, some implications are considered.

Autism Spectrum Conditions

Autism is usually presented as a severe developmental disorder. The term was coined by Kanner (1943), and was intended to convey the manner in which those with the condition tended to withdraw, in particular from social interaction. A reluctance to engage in, or inability to deal successfully with, social interaction is the major feature focused upon in both describing and attempting to explain the condition. Such attempts have however been complicated by the fact that the manifestation of the condition varies dramatically across different individuals. Recently this variability has been captured by talking of a spectrum of conditions, more specifically of autism spectrum conditions,[2] including classic and high-functioning autism and Asperger's Syndrome.[3] Differences along the spectrum relate to the severity of the symptoms but also mark some quite different characteristics, of which the aptitude for language use is perhaps the most striking. For the remainder of this chapter, I shall use the term 'autism' to refer to this more general spectrum of conditions rather than, as is sometimes the case, to the more extreme end of it.

Whilst this variation of characteristics makes it difficult to talk consistently of a well-defined problem, a concern with the term autism has significantly grown in popularity in recent years. There seem to be several reasons for this. First, it is generally agreed that whatever the precise cause or particular manifestation of the condition, a common denominator is that there is some kind of (often

---

[2] Much recent literature actually uses the phrase 'Autism Spectrum Disorder'. The problem with using the term disorder here is that it suggests the idea that there is some fault or disease that can be 'cured'.

[3] Most recently there has been some debate about whether there should be continued use of the term 'Asperger's'. Although there is clear continued use of the term amongst practitioners.

severe) impairment[4] of social behaviour and social relatedness, or simply a failure to understand and to deal appropriately with social interaction (see DeMyer et al., 1981). Another reason is that it has been convincingly argued that the idea of a spectrum of conditions suggests that classic autism can be seen as an extreme version of characteristics that are actually very common, in lesser forms, in much of the rest of the population. In particular the understanding of autism as a manifestation of an 'extreme male brain' has piqued the interest of not only academic researchers, but much of the general population (Baron-Cohen, 2004).[5] Recent conceptions of autism have made it possible to reconsider many (especially male) character-istics that have in the past been interpreted positively, as milder versions of characteristics that are now considered to generate severe problems if taken to excess.

The other aspect of autism research that has generated keen interest is the idea that autism is at root genetic, detectable as particular atypical development in the structure of the brain (Bauman and Kemper, 2005, Neuhaus et al., 2010, Fatemi, 2016). Various lines of research are currently being pursued which seem to provide wide-ranging, and mutually reinforcing, evidence for such ideas. For example, there is evidence for the role of genetics, although with a role for environmental factors as well (Stodgell et al., 2001, Geschwind, 2008). Others have found evidence to suggest that neural development is hampered by inflammation in the brain (Young et al., 2016, Vargas et al., 2005), neuron density (Courchesne et al., 2011), fetal testosterone (Baron-Cohen et al., 2011), the atypical pruning of synapses (Geschwind and Levitt, 2007), right-side left-side imbalances of the brain (Floris et al., 2016) and dysfunction in the prefrontal cortex (Cooper et al., 2016, Shalom, 2009, Uddin, 2011). Many of these factors, such as fetal testosterone, are not only crucial developmental factors but also

---

[4] One further note of caution at this stage – although the terms 'suffer' and 'impairment' and indeed 'disability' are frequently used to describe autism, it is clear that there is growing opinion that autism should not be seen this way at all. And, for example, that many of those described as autistic would not wish to be 'cured' even if such a possibility existed (see especially Grandin, 2006). For an excellent discussion of this issue see Murray (2011).

[5] It is worth mentioning also that Asperger originally believed (for some seven years), that the condition he described in 1944 was a variant of male intelligence and that women could not have the condition.

would account for the high prevalence of autism amongst males (Baron-Cohen et al., 2015).

Despite the hive of activity that has focused upon the neurophysiological underpinnings or causes of autism, there remains fairly consistent agreement about the main ways in which the condition is manifest. Central to autism is the idea that the social world is difficult to cope with. Ever since Kanner's original description of autism, theoretical contributions have made a very strict distinction between relations to the material world and those to the social world (Williams et al., 1999). In Kanner's early work, a very clear contrast is drawn between the 'fascination and *excellent* relations with objects' that autistic children display, as opposed to their disinterest and non-existent relations to people and anything usually termed 'social'. Subsequent accounts of autism have, for the most part, left such observations unchallenged and have even made a more explicit and pronounced separation of the natural and social worlds (Baron-Cohen et al., 1985, Leslie, 1987, Rogers and Dilalla, 1991, Hobson, 1993). In short, children with autism are considered to be relatively successful at negotiating the physical world of things, but experience severe problems with the sociocultural realm of human interaction.

The idea that those with autism experience difficulties in negotiating the social world is supported by a wealth of empirical evidence. Several studies have shown how those with autism have difficulty understanding things that are known or believed by others (Baron-Cohen et al., 1985, Perner et al., 2002). Those with autism are portrayed as having problems identifying the meaning of facial expressions, matching them to appropriate gestures (Weeks and Hobson, 1987) or anticipating the kinds of things that a listener needs to know (Baltaxe, 1977, Loveland et al., 1990, Mundy et al., 1986, Wulff, 1985). Often such factors are put together and talked about in terms of the trouble those with autism have detecting the meaning of social situations (Loveland and Tunali, 1991, Lawson et al., 2004). To repeat, it is this difficulty with negotiating the social world that is the major phenomenon that theories of autism seek to explain.

Theories of Autism

The three most prominent theories of autism are the 'theory of mind', 'weak central coherence' and 'weak executive function' theories. By far the most prominent and influential of these is the theory of mind approach, which has largely been associated with the work of Simon Baron-Cohen (Baron-Cohen et al., 1985, Leslie, 1987, Baron-Cohen, 2016, Baron-Cohen et al., 2015). The main idea, also sometimes captured with the term 'mind-blindness', is that those with autism are thought not to be able to recognise that others have minds and, it is supposed, it is this that makes social interaction difficult.

A series of studies demonstrate (albeit it in high-functioning, verbally capable autistic children) that there is a tendency for those with autism to make far greater errors in predicting the knowledge or beliefs of others in a variety of different situations. A crucial part of this theory is the idea that those with autism have difficulties attributing false beliefs to others (Frith and Happé, 1994).[6] The implications drawn from such studies are that there is a significant 'decoupling' of mental representations from reality. Although such decoupling has problems for all kinds of developmental activities, the main concern of theory of mind approaches is with the relation of the person with autism to others. The capacity for meta-representation is, they argue, essential to anything other than the most superficial understanding of 'other minds'. Other people may believe, think or say things that are in some way at odds with reality – they may be mistaken or lying. The child must be able to uncouple their representations of what the other says, thinks, etc., from reality.[7] Autism, on this approach, derives from a difficulty in the decoupling that is required for the emergence of meta-representations. Such decoupling, it is suggested, is critical for social interaction, but incidental to dealing with the natural or material worlds.[8]

More recently these ideas have been recast in terms of a distinction between empathising and systematising (Baron-Cohen et al., 2002,

---

[6] The most famous 'false-belief' test is the Sally Anne test. Here Sally, after playing with a marble and leaving it in a particular place, such as a basket, leaves the room. Anne then moves the marble to another location and the subject is asked where Sally will look for the marble. Children with no autism recognise that Sally does not have the information that the subject has and so will look in the basket. Those with autism tend to reply that that Sally will look in the new location, so not realising that Sally can have a false belief (Baron-Cohen 1995, p. 71).

[7] Thus the precise meaning of meta-representation is not simply that of a higher-order representation, although there is clearly some dispute about the consistency of the term as used in autistic research (Scott, 2001).

[8] This distinction is usually made in the context of a discussion of 'pretend' and 'functional' or 'reality' play (Leslie, 1987).

Baron-Cohen, 2004). A theory of mind is still central, but the emphasis upon empathising signifies not only problems with identifying other emotions, feelings and thoughts, but with the ability to respond appropriately. Systematising refers to the inclination to analyse and build systems with the aim of understanding or predicting (typically) non-human events or mechanisms (Baron-Cohen, 2009). The idea of a 'typically male brain' is then introduced which, it is argued is more attuned to 'systematising' tasks, such as creating and modifying taxonomies. Similarly, the idea of a typical female brain is also introduced which is, in comparison, more attuned to 'empathising' tasks, such as learning what others are thinking or feeling by reading the language of facial expressions (especially the 'language of the eyes') as well as the ability to respond spontaneously in an emotionally appropriate way. These abilities stand in contrast to one another, with strengths in one appearing alongside deficits in the other (Happé, 2000). On this account, those with autism are understood to have an extreme version of the male brain. The inability to empathise is viewed as accounting for the noted unsociability and autistic aloneness (Barnbaum, 2008). Whatever one might think about the idea of a typically male or female brain, it is clear that autism is not evenly distributed amongst men and women and that in providing an explanation in terms of a typical male brain, Baron-Cohen and others are attempting to capture and explain the significantly higher prevalence, especially of Asperger's syndrome, in males.[9]

The second theory, that of weak central coherence, focuses on the idea that those with autism fail, in effect, to see the bigger picture, rather focusing upon the details of some situation or state of affairs that are irrelevant to the significance of the whole. The term 'central coherence' refers to the ability to draw together details so as to recognise the meaning of the entire picture. If this is in some way impaired, then it becomes difficult to make generalisations or construct mental paradigms (Snyder, 1998). In this case, the world is full of surprises; everything has to be examined anew by treating each detail with equal importance. Thus impairments in social interaction, deficits in communication and repetitive behaviours are all explained in terms of attention to detail to the detriment of the whole.

---

[9] Although there is some evidence that girls may be better at camouflaging their autism, and thus biasing the data for this, the evidence for a preponderance of autism amongst males is still overwhelming (Rynkiewicz et al., 2016).

The third theory is that of weak executive function. The argument here is that this capacity, executive function, facilitates planning and organisation (Perner and Lang, 2000) and is crucial for decision-making, keeping several tasks going at the same time, resolving conflicting responses and inhibiting inappropriate behaviour (Frith, 2003). Where there are problems with such tasks, weak executive function is understood to underlie failures to update information. Several tests have been used to identify poor executive control function, including a variety of card sorting tasks that require the participant to sort the cards based on colour, shape, category, etc.[10] The rule for sorting that will be used during the test is not disclosed to the participant. Instead, the participant must work it out from the experimenter's responses to the participant's choices of cards. For example, a set of cards may have animals that are coloured either red or blue. If the rule the experimenter is using is based on colour, and the participant works this out, the experimenter may change the rule to types of animals. In order to succeed, the participant must become aware of this rule change and alter their responses accordingly. Those with autism tend to fare badly at such tests, unable to let go of a rule once they have worked it out.[11]

There has been much discussion in the autism literature of the degree to which these three theories compete with each other, and a certain amount of agreement that there is huge scope for overlap between them (O'Loughlin and Thagard, 2000, Gluer and Pagin, 2003, Happé, 2000, Frith, 2003, Barnbaum, 2008). One attempt to combine the three theories is provided by John Lawson (2003).[12] Lawson argues that each theory is a subset of a larger thesis, which he terms the Depth Accessibility Difficulty model. Here the emphasis is still upon cognitive capabilities, but these are understood in relation to particular theories of being or ontologies.

More specifically, Lawson takes as his point of departure the main ontological commitments outlined in Chapter 3. In particular, Lawson makes a distinction between an ontology of constant conjunctions of

[10] For example, the Wisconsin Card Sorting Task.

[11] More directly, some tests simply involve experiments where a sponge is made to look like a rock, and children are told that it is a rock, those with autism will continue to believe that it is a rock even after it is shown to be a sponge (Nichols and Stich, 2000). Thus, first impressions last.

[12] Another fascinating account is provided by Barnbaum who argues that all three theories involve the idea that those with autism have some kind of inability to make intentional state attributions (Barnbaum, 2008).

atomistic events and one of powers, mechanisms and structures irreducible to events and states of affairs. His main idea is that those with autism have difficulty viewing or comprehending the world in any terms other than those of closed systems of atomistic events. Those aspects of the world that most conform to such closed systems are then not difficult to comprehend, engage or feel comfortable with. Indeed, he argues, it is only in relation to these aspects of the world that those considered as high functioning display their often extraordinary talents or abilities. However, those aspects of the world which are irreducibly 'open', as much of the social world seems to be, present more of a challenge (J. Lawson, 2003). In other words, in the terms used above, those with autism have only limited use of methods of abstraction, but have relatively few problems with methods of isolation.

As noted, Lawson motivates this discussion by attempting to locate existing conceptions of autism within a general model. Theory of mind theorists, focus upon a deficiency of those with autism in understanding feeling, emotions, etc., but at the same time acknowledge their abilities in building or analysing mechanical systems. Now if there is some problem with recognising or coping with depth, underlying needs, motives, beliefs, etc., will just not be accessible or understandable. Understandings of other people, especially their beliefs and desires and the role that these play in their causal or structural capabilities, will tend to be reduced to the actions that can be observed. And for generalisations of behaviour to be possible, which of course is crucial to the negotiation of social reality on a daily basis, this will only be possible by adhering to, and hoping that others adhere to, strict rules of the kind 'whenever X then Y'. The nuances of social interaction such as thinking of what some other person will *tend* to do, coping with contingencies, understanding how rules change in different contexts, and how they can be combined in non-atomistic ways will tend to be challenging or will be missed by those with autism.[13]

More generally, the kinds of underlying mechanisms that serve to structure social reality, given their dependence upon human activity, community agreement and meaning will, as noted above, have a different mode of existence from the kinds of natural mechanisms

---

[13] In these instances, such individuals will engage with the capabilities they have access to. Lawson terms such ways of seeing the world as 'closed system thinking' or 'misclosure' (J. Lawson, 2003).

that tend to be the object of experimental activity. Mechanisms and powers in the social world will tend to be inherently processual and relational in character. More specifically, if much of social reality is indeed processual and relational, then those with autism will tend to find it difficult to cope with changes in rules and fashions that are ambiguous, context dependent, negotiable and continually under transformation. Similarly, if much of social reality is relational, the idea that people (or objects) occupy different, and multiple, positions and roles will be difficult to comprehend, as will the different kinds of emergent systems discussed in Chapter 3. In this way the concerns of the executive function and the weak central coherence theorists, as well as those of the theory of mind, are all accommodated within a more general framework. Moreover, the observed greater competency with mechanical and atomistic systems is also explained (J. Lawson et al., 2004).[14]

It would also seem that more recent accounts of autism are starting to investigate just these issues. For example, Peter Vermeulen has recently attempted to accommodate the same observations about autism (especially those usually focused upon in mind-blindness accounts) in terms of what he terms context blindness (Vermeulen, 2012). As with J. Lawson, Vermeulen's main point is that the key to understanding those with autism lies not in what they do but how they think about the world. But for Vermeulen, the problem lies with the ability to deal with

---

[14] Although Lawson does not formulate things in these terms, it may be possible to understand how those with autism may be acting in conditions where isolatability is not present (i.e. where it is not possible to treat reality as a kind of closed system) in terms of what Bhaskar has termed the recourses open to actualism (Bhaskar, 1978, p. 67). Briefly put, if in some situation isolatability is not present, and so closure not possible, then it will be impossible to capture or refer to reality in terms of simple 'if X then Y' rules or laws. But if some individual persists with a conception of reality in terms of constant conjunctions of events, then two courses of action are open, both of which would seem only to lead to infinite regresses. On the one hand, there is a tendency to include more and more in their model of the world with the result of considering systems so vast and unwieldy that nothing is excluded. On the other hand, there is a tendency to dig down to smaller and smaller elements in the hope of finding some atomistic elements that always behave the same way under the same circumstances, with the result of considering systems so small (and unchanging) that they actually include nothing. This seems to capture something of the documented evidence that those with autism tend to amass unusable amounts of information or become obsessively preoccupied with some, often considered irrelevant, detail.

what he terms 'context'. However, on closer inspection, it seems that the main features of Vermeulen's context are just those aspects of social interactions that underlie the interconnectedness of social phenomena and the openness of the systems within which they take place, identified by Lawson. Thus those with autism, at least the high functioning ones, know many rules of social interaction, but fail to deal successfully with the changing contexts in which they might apply or with exceptions to those rules, which come about because of the interrelatedness of different contexts. Thus it is not that emotions of others cannot be fathomed or empathy generated. Rather it is a lack of contextual sensitivity to the ambiguity of the stimuli faced that causes the problems. In terms of the above, there is an overreliance on 'whenever X then Y' statements, irrespective of the conditions that prevail, or a failure to abstract when confronted with open systems.

Autism and Technology

Given the account of technology provided in earlier chapters, there are some fairly obvious reasons for suspecting that those with autism will tend to be drawn to certain aspects of technology. However, before proceeding to spell these out, I first consider existing evidence that some kind of special relationship tends to exist between those with autism and technology. There appear to be many different sources of such evidence, but for present purposes I shall restrict attention to those noted at the outset, focusing in particular on the importance of assistive technology, recent reports of autism 'epidemics' and research into genetic factors affecting autism.

Recent years have seen a growing awareness and support for the ways in which technology might be used to help assist in the education and treatment of autistic children (Wainer and Ingersoll, 2011). And as noted, there is a stock of anecdotal evidence that children with autism are attracted or drawn to technological devices, which has led researchers to devise treatments that take advantage of this fascination (Colby, 1973). The perception that technology has much to offer in this regard has even led to the emergence of several new journals in the area.[15]

---

[15] For example, the *Journal of Special Education Technology*, the *Journal of Educational Multimedia and Hypermedia*, the *Journal of Computer Assisted Learning*, etc.,). In addition, clinical psychology journals are recognising the

Different kinds of technology have been employed for a range of tasks such as mechanical prompting (Taylor et al., 2004), the use of video for instruction (Sturmey, 2003, Bellini and Akullian, 2007), virtual reality for the practicing of interaction (Max and Burke, 1997), computer-based interventions aimed at improving social problem-solving skills (Bernard-Opitz et al., 2001), vocabulary (Moore and Calvert, 2000) and vocal imitation (Bernard-Opitz et al., 2000). One particularly interesting case has involved the use of robots as therapeutic toys or playmates. In the AuRoRa Project, it is claimed that the mechanistic qualities of robots, which some view negatively, are the very features which make robots the preferred partners for children with autism (Robins et al., 2012). For example, truck-style robots were designed in the hope that a non-human appearance might facilitate interaction (Graham-Rowe, 2002) and teach basic social skills encouraging more capable social behaviour (Dautenhahn, 2003). Common themes run throughout these contributions. In particular, the essentially mechanical features of technological devices are more attractive or less frightening than social interaction.

The second source of support for a special relationship between those with autism and technology has emerged in attempting to explain what some have termed an autism 'epidemics' in recent years. On the face of it, the evidence is overwhelming. It has been calculated by the US Department of Education that for the ten-year period, 1993–2003, there was a 657% increase in nationwide rate of autism.[16] In the first two decades of the twenty-first century, numbers have risen from 1 in 150 children being affected in 2007, to 1 in 68 in 2016.[17] Of course, there has been significant debate about whether such increases are really due to increases in the prevalence of autism or due to changing methods of detection. And given that the condition manifests as a spectrum of different characteristics, it is inevitably difficult to specify clear boundaries to either having or not having the condition.

Whether or not the overall prevalence of autism has increased, there is no disputing the geographically skewed nature of the prevalence of autism. In particular, there are massive concentrations of autism in

importance of technology in service delivery and devoting special issues to the topic (e.g. Newman, 2004).

[16] See, for example, *Scientific American*, Dec 2007, 17, 58–61.

[17] These figures taken from the Centre for Disease Control and Prevention: www.cdc.gov/ncbddd/autism/addm.html.

predominantly high-technology areas, the most famous being Silicon Valley and Route 128 outside Boston in the United States.[18] Many accounts now exist of such concentrations, but perhaps the best summary is still that provided by Silberman (2000).

He notes a wide variety of different kinds of indicators to this effect, such as Microsoft's offer to employees of insurance benefits to cover the costs of behavioural training for autistic children as standard in areas where technology industry is highly concentrated. Elsewhere, he notes simply that 'at clinics and schools in Silicon Valley, the observation that most parents of autistic kids are engineers and programmers is not news'.

A crucial part of Silberman's account is his focus upon the reasons why those with (mild) autism are likely to be very drawn to engineering jobs, especially computer programming in such high-technology areas. All programming requires a little autism, Silberman suggests. Multitasking is often difficult for those with autism, especially if one of the tasks involved is face-to-face communication. Replacing the social 'hubbub' of a traditional office with a screen and email, inserts a controllable interface between a programmer and the everyday chaos of the workplace. Further, the flattened hierarchies of many software firms are more comfortable for those who have trouble with many social clues. Moreover, those who have been successful in information technology industries, such as Bill Gates and Mark Zuckerberg are often portrayed as 'on the spectrum'; with Zuckerberg in particular often being described as 'brilliant with technology but pathologically bereft of social graces' (ibid.). Moreover, programming is often represented as requiring long attention spans, careful repetitive tasks, painfully close attention to detail, etc., all characteristics highlighted as present for those with high-functioning autism.

But whilst all this may explain why those with autism are drawn to the high-technology areas noted, it does not of itself explain the rising prevalence of autism in children. The most common explanation of this is that, in the past, those with autism will have tended to live out their lives in rather isolated ways, making few friends and standing little chance of finding a partner. However, high-technology hubs are attracting large numbers of similar individuals. Engineers, scientists and

---

[18] Recent studies have found similar clustering in less well-known areas such as Eindhoven in the Netherlands (Roelfsema et al., 2012).

programmers are likely to meet compatible partners at work. It also seems to be the case that parents and grandparents with very mild forms of autism are likely to have offspring with much more pronounced characteristics,[19] and that this is particularly evident with parents interested in technology (Wheelwright and Baron-Cohen, 2001).

Whilst such evidence suggests some kind of special relationship with technology, and in particular a heightened proclivity to technology-related capabilities, it says nothing about which aspects of technology are involved. Kanner's initial idea, which has gone largely unchallenged, was that those affected would have 'excellent relations' with physical objects. But many, such as Silberman, suggest that those with autism are actually physically clumsy and overwhelmed in much of the physical world, suggesting that the attraction to technology is really motivated by an attraction to the *virtual* realms of mathematics, symbols and code (Baron-Cohen, 2016). However, if the DAD thesis is correct, the crucial issue is the relative importance of closed (or closable) systems, even though knowledge is typically gained in open-system situations.[20]

To pursue these questions, it is helpful to look a little more closely at the relationship between those with autism and physical objects. Indeed, when many write of the relationship between autism and technology it would seem that their conception of technology is one that in fact goes little beyond that of material objects or devices (e.g. Goldsmith and LeBlanc, 2004).

---

[19] This of course presupposes that there is some kind of genetic component to autism. But evidence for this seems now generally accepted. More specifically, there now seems good evidence for the following propositions. First, that there is at least some kind of genetic component to autism, given that if one twin is autistic, there is a 90% chance that the other is, and that if one child is autistic, the risk of the next goes from 1 in 500 to 1 in 20, and then to 1 in 3.

[20] Lawson points out that those with autism are likely to develop better capabilities for dealing with systems that can be understood atomistically and/or mechanistically. But the physical world is of course just as structured and just as out of phase as the social world. And although experimental controls may only be possible in relation to the mechanisms of the natural world, most of our daily interaction with physical objects does not actually take the form of constructing experimental conditions and setting about triggering one kind of mechanism or another. In short, much of our knowledge of the physical world is acquired in open systems too.

Material Objects and Affordances

A range of extensive, if fragmented, evidence exists concerning the use of material objects by those with autism. A brief review of some of their main findings is as follows. First, focusing upon the forms of exploration of objects that those with autism develop in early years, there appears to be a persistence with what is usually termed 'undifferentiated manipulations' of objects (waving, banging and mouthing). There seems also to be a preference for exploring with proximal senses of touch, taste and smell (Sigman et al., 1983, Freeman et al., 1984, Adrien et al., 1987, Libby et al., 1998). Moreover, visual inspection of objects frequently takes an unusual form, where objects are twisted and viewed from unusual perspectives (Wing, 1974, Hermelin, 1970). Another recurrent feature is that just one object or single part of it is looked at for extended periods of time (Gillberg et al., 1990). Objects are also used in mechanical ways. For example, they are lined up in rows or on top of one another (Sigman and Mundy, 1989).

A second set of findings relates to the problems those with autism seem to have with learning about objects by interacting with others. For example those with autism rarely engage in 'pretend play' and have great difficulty sharing objects with others or learning about an object or its use through copying or interacting with others (Sigman et al., 1992). They seldom engage in social referencing. Any imitation of how others use objects takes place in a very mechanical and inflexible manner (Smith and Bryson, 1994). Moreover, those with autism tend to be fairly oblivious to the status of objects, for example as property. Not only are they indifferent to other's property, but to many of the objects that rightly belong to them (Bosch, 1970, Zabel and Zabel, 1982).

A third set of findings relate to the idea of there being a conventional or appropriate way of dealing with objects. Those with autism tend to pick out isolated aspects of an object regardless of its conventional use, and fixate on features of objects which are often thought irrelevant or uninteresting and not normally noticed by others (Lord et al., 1983, Freeman et al., 1984). Frequently cited examples are preoccupations with shapes, such as brickwork (Bosch, 1970) or the shape of a jigsaw puzzle rather than its picture (Frith, 1989). They also tend to categorise or collect objects on the basis of a peculiar feature, regardless of their conventional uses (Wulff, 1985).

Williams and Costall, having reviewed much of this evidence, conclude that those with autism have problems with objects in two interconnected ways. First, they are slow to understand the conventions concerning their use and secondly, they tend to make sense of objects and explore them in very idiosyncratic ways, which will often seem bizarre to those around them.

Such ideas have been developed in terms of (especially social) affordances (Loveland, 1991). Affordances are usually interpreted as the opportunities for perception and action offered by the environment (Gibson, 1977, 1986).[21] Significantly, in Gibson's account, affordances are perceived directly rather than inferred, deduced or retrieved by association; they capture the idea that what we attend to in our surroundings are not simply properties such as shapes and edges, but the opportunities things offer up for perception and action relative to the specific physical and psychological capacities of the particular organism. For example, food is 'edible', a hand is 'shakeable' a chair is 'sit-on-able'. The affordance of sit-on-able-ness is only an affordance of a chair for animals with a flexible torso, hip joints, legs, knees of appropriate heights, etc. Thus in a way, the focus is upon the way that affordances link the perceiver to situated action or to their environment.[22]

---

[21] This idea has been applied to a range of psychological problem areas (e.g. Baron and Boudreau, 1987, Heft, 1989, McArthur and Baron, 1983, Smith and Ginsburg, 1989, Loveland, 1991, Williams et al., 1999). However, it is worth pointing out that those drawing upon affordances ideas have been motivated by a variety of dissatisfactions with the prevailing sate of psychology – especially its lack of emphasis upon the structure and informativeness of the social environment and neglect of the role of culture in cognition (Valenti and Gold, 1991). In other words, there has been a reaction to the exclusive concern with internal cognitive structures, with little regard for the specific material conditions in which they might operate, and it is the latter that are of *particular interest here.*

[22] However, these properties are objectively given irrespective of whether they are perceived or not. Indeed, the idea is similar to earlier Gestalt ideas, e.g. to the 'demand character' of objects – a postbox, invites the mailing of a letter because its meaning is immediately apprehended, 'as sure as its colour or form, whenever a person needs to mail a letter' (Koffka, 1935). But affordances do not change as the needs of the observer change, as for Koffka. Rather 'an affordance is not bestowed upon an object by a need of an observer and his act of perceiving it. The object offers what it does because of what it is' (Gibson 1986, 139). But differences in perception of affordances result from anatomical differences and from the behavioural niche occupied by the organism.

However, Gibson's primary concern is with directly perceivable features of the lived environment. Here, direct perception derives from the environmental surfaces that 'structure ambient light in lawful and specific ways, and because the visual systems of humans have evolved in a way that permits the extraction of these lawfully generated patterns of ambient light' (quoted in Valenti and Gold, 1991). This is the strength of the affordances approach: it focuses in a very detailed way on the manner in which information is conveyed about the world in terms of ambient light, sound and odour 'and other information obtainable by mechanical means' (Gibson, 1977).

On reading through extended accounts of affordance applications it is clear that these more 'mechanical means' involve various conditions (see especially Gibson and Pick, 1979). Objects are discrete objects, with clear boundaries that are both easily differentiated and articulated. In particular, physical objects behave the same ways in the same circumstances. Thus although learning does not take place in the kinds of closed systems achieved in science, knowledge of affordances is still typically of the kind 'if X then Y' kind. In fact a way of summing up much of the affordance idea is that it concerns the various complex ways in which positioned or situated knowers come to have knowledge of the causal or intrinsic powers of structured things without constructing experimental controls (see Gibson and Pick, 1979, especially Chapter 6).[23]

---

[23] It is, however, worth pointing out that the means are very different to those of experimental activity. For example, it is notable that although the affordance approach emphasises the importance of exploration and investigation in coming to know about the affordances of things, the kinds of exploration considered are either actually only more sophisticated forms of viewing (such as viewing from a different perspective):

Information must be actively sought: it does not fall, like rain, on passive receptor surfaces. We move our heads to disocclude a portion of a temporarily invisible scene; we step forward to magnify a wanted view; we lean to peer around a corner or glance in our rear-view mirror as we back the car. (Gibson and Pick, 1979, 20)

   Or the acquisitions of skills:

'A three-inch-wide beam affords performing backflips for a gymnast, but the affordance is not realizable by others; rock climbers learn to use certain terrains for support that do not appear to others to provide a surface of support' (ibid. 17).

There seems to be no place for the kind of active intervention in the world that is central to creating the closed systems that have such significance in natural science, even on a partial level.

In this context, Loveland's contribution is to distinguish three kinds of affordances in human environments.[24] First, she distinguishes affordances for physical transactions with the environment (sitting on, walking, eating, etc.,). Such affordances are drawn upon in dealing with the immediate physical environment – with digging, picking up, manipulating, etc. Second, there are culturally selected affordances that reflect preferred but not necessary interactions. For example, socks afford wearing on one's feet. Although they can be used as containers for small objects, such as cutlery, these are not their 'preferred' or standard use. Affordances apply to natural objects as well as artefacts. But preferred affordances are culturally selected (Reed, 1988). In other words, preferred affordances reflect participation with others in a shared cultural milieu that predisposes the individual to use objects and interpret events in certain ways. Part of being creative, Loveland argues, involves the ability to transcend the preferred set of affordances in appropriate circumstances. The third set of affordances, are social and communicative, reflecting the meaning of human activity for others. These include symbolic behaviour such as conversation and writing, but also gesture or facial expression, body posture, tone of voice, direction of gaze, etc.

Loveland concludes that the evidence collated by Williams and Costall shows that autism involves an impairment or difficulty with the latter two kinds of affordances. Specifically, autism involves a specific impairment in the child's ability to detect not only the affordances of other people, but culturally selected or agreed upon affordances of things. In the language of Chapter 6 it is tempting to summarise Loveland's position as saying that those with autism have difficulties understanding the workings of icons and social objects, but are more comfortable with tools and technological artefacts. It should be clear that in all of the above there is every indication that those with autism are both drawn to and more comfortable with those objects for which intrinsic causal capabilities are most important or for which intrinsic capabilities exhaust what is needed to be understood for capable or successful use in a particular context.

---

[24] Which are also layered – in the sense that any object, person or event, simultaneously presents multiple kinds of levels of meaning to the human perceiver.

Just as important, but less obvious, is the idea that isolatability seems to lie at the heart of the different ways of coming to know and abilities to negotiate different aspects of reality. Indeed, this discussion of Williams and Costall's results would seem to suggest that rather than talk about natural, physical or material versus social domains or mechanisms, it may be better to talk of different domains or spheres of differentiability, where the term differentiability refers to the extent to which different kinds of mechanisms in reality can be isolated and yet still work in roughly the same ways. It is, of course, this differentiability that underlies the success of controlled experiments. Where differentiability is present, such as under successful experimental conditions or in the direct apprehension of the causal properties of material objects, those with autism will have few problems either understanding or acting appropriately. Where such differentiability is not possible, such as in the midst of inherently processual and internally related networks of social interaction, those with autism will be more challenged. In these latter situations, often the best that can be expected for those with autism will be the often rather inappropriate imposition of 'if X then Y' type reasoning and idiosyncratic simplifications (isolations), often becoming preoccupied with details that will seem irrelevant to many.

On the basis of the above, the main point to make about physical objects is that they can be comprised of both social and physical constituents and that we would expect those with autism to not simply 'perform excellently with physical objects' as Kanner suggested, but to display a wide variety of capabilities, in part dependent upon the degree of differentiability of the object being used, its context and the kinds of affordances being focused upon.

This attraction to isolatability is particularly evident in the exploratory activity of autistic children with material objects that Williams and Costall focus upon. Here the kinds of exploration of the affordances of things take a particularly mechanistic form, relying on forms of isolation or inappropriate abstraction. Single details are fixated upon whilst ignoring others. Another recurrent feature, focused upon in particular by Loveland, is the inability of those with autism to understand the positioning of objects in social networks of use or play. Whereas much of social activity involves the reproduction or transformation of social relations, conventions, norms, etc., in networks of interaction, this is problematic for those with autism. Whether the problem is working out normal roles (locations for such objects within the system or network) or particular relational characteristics – such as private or personal property – there appear to be severe constraints to understanding what is happening or expected.

Autism, Technology and Isolation

If this account is correct, perhaps the central problem facing those with autism is that it is simply impossible to isolate the bulk of the affordances of the social world. This explains not only why some aspects of their day-to-day lives are difficult, but why other aspects, where such isolatability is indeed possible, might be particularly rewarding and comforting. In this context, it is possible to start to construct an account that explains the particular relationship between those with autism and technology noted above.

The important aspects of the above account of technology, for current purposes, are the following. Technology can be understood synchronically, in terms of the harnessing of the causal powers or capacities of technological artefacts in order to extend human capabilities, and diachronically, in terms of the moments of isolation, recombination and enrolment. At the heart of this account is the combination of social and non-social factors, which manifest themselves in terms of the relative isolatability of the causal powers harnessed in any particular technical activity.

On this account of technology, it should really be no surprise that there is some kind of special relationship between those with autism and technology. A central feature of technology is the intrinsic causal powers of well-defined 'things'. As the bulk of such powers, or affordances, are relatively isolatable, in the sense that much of the natural world is conducive to controlled experiment, then there is much about technology that is amenable to the kind of 'closed-system' thinking (methods of isolation) that seems to best capture all the broad features of autism. Moreover, it is easy to understand which particular features of technology are most likely to be of interest, or of comfort, to those with autism. In particular, it should be no surprise that those with autism are most drawn to the isolative and recombinatory moments of technology. Thus, underlying the observations of 'epidemics' of autism in high-technology areas, is the fact that those with autism are attracted to the design stage of technology, for example, where obsessive concern with details and repetitive tasks, as evidenced in coding, is deemed essential. Moreover, as the affordances literature suggests, the social positioning or enrolment of technology is something that causes far more challenges for those with autism, often taking place in rather idiosyncratic and unexpected ways. Whilst this may take place in ways which are innovative as well as unexpected, it is unlikely to be something that comes easily or even bears much relationship to collectively agreed uses and behaviours.

Stated in these ways, the account of technology developed above and the Depth Accessibility Deficit thesis of autism, may be seen to support each other. If the account of autism is correct, then the kind of problems that those with autism have with the social world underlie there being important distinctions to be made between the social and the non-social, as suggested in the account of technology given above. Alternatively, if technological artefacts can be understood as suggested, it does seem possible to account for the general evidence that there is a special relationship of those with autism and technology, whilst those with autism tend to be more challenged by the positioning of technological artefacts.

Moreover, if both accounts are correct, a series of implications follow. Given space, it is only really possible to comment briefly upon three quite different examples. The first concerns the implications for ideas concerning the way in which 'dual-nature' theories of technology can be understood, and in particular for the idea that the social and non-social components of technology become mangled together in practice. The second concerns possible worries about future trends for technological design. In particular, it seems to open up further grounds for considering existing ideas about the democratisation of technology. The third relates to grounds for extending feminist theories of technology. I shall briefly outline each, in turn.

Un-mangling the Social?

In Chapter 3 I focused upon one possible criticism that might be made of the account of technology developed in this book, that it is in practice impossible to make any useful distinction between the social and non-social, focussing in particular on the work by Latour. Here I want to focus upon a related argument, which was touched upon in the discussion of sociomateriality encountered in Chapter 5 and can be found in recent discussions of new materialism,[25] but which is perhaps brought out most clearly in the work of Andrew Pickering (especially in Pickering, 1995). Although Pickering does accept that analytical distinctions between social and non-social factors are required to make sense of technology, such factors become effectively 'mangled' together in the ongoing and everyday performance or 'practice' of engaging with technology. Thus, whatever ontological differences there might be between the social and non-social these have no significance in practice as they become so entwined as to make any form of separation impossible (Pickering, 1995).

Although keen to establish a different position to that of actor network theory, Pickering's position actually draws quite heavily from it and is certainly in agreement with Latour that there is a 'constitutive intertwining between material and human agency'. For Pickering, who is approaching the issue from the perspective of giving an account of science, rather than technology, humans must be understood as located within 'a field of material agency that they continually struggle to capture (or harness) in machines' (Pickering, 1995). Pickering frames such ideas in terms of a 'dialectic of resistance and accommodation' as human agents devise different methods to capture material agency under different conditions.

The two main points for Pickering are that genuine temporal emergence makes it impossible to predict what will happen next, and that the human and non-human are inextricably entangled in that emergence (grounding, for Pickering, a particular form of post-humanism, ibid, p.23). This is the mangle of practice – in which the social and non-social (or human and material agency) are constitutively entwined. The main implication is that it is pointless to attempt any analysis based on some generalised distinction between the social and non-social. The argument is slightly different from that of Latour. As noted above, Latour's unwillingness to talk of the social rests on the ideas that there no separate zones of the social and material, no social forces and no social material – there are only ontologically flat 'happenings' where a range of actants cause and are caused, giving the social scientist little else to do other than trace (describe) the nature of their associations between a range of different kinds of actants. For Pickering, the problem lies in the complexity of the social-material

---

[25] For example, see Coole and Frost (2010), Connolly (2013) and Coole (2013).

distinction (making it practically impossible to make any general statements about social or non-social domains).

However plausible Pickering's position sounds, the above account of autism and technology does seem to suggest that the social is perhaps not in practice quite as 'mangled' as Pickering suggests. I must be careful about what I am claiming at this point. In particular, I must be clear about the relationship between what I have been terming 'the social' and what I have been terming the isolatable, and my suggestion that there is some kind of link between the two. A major point here is the finding of social ontology noted in Chapter 3, in particular that the social world is not isolatable in the same way as much of the non-social world. Returning to the idea that phenomena can be understood in terms of particular forms of organisation, not all kinds of organisation are the same. A central distinction, given this emphasis upon process and emergence, is that some groupings of factors are likely to be more able to 'stand by themselves'. Thus in much of the natural world, although 'things' are made up of organisations of lower level 'elements', the resulting organisations seem to be relatively well defined and individuated causal agents. As noted, there is nothing a priori about such distinctions, but it does seem to be the case that, given the prominent role of replicable experiments in natural science but not in social science, such distinctions are significant. Indeed, this is the main task of experiments, to isolate fairly well-defined and coherent sets of mechanisms that operate in pretty much the same way in the same circumstances. Such isolatability, however, which seems to be at best 'partial' in the non-social world, is likely to be very rare indeed in the social world, very much as Pickering suggests, because it is an emergent feature of the *interaction* of human agents. Thus, any social causal powers are the emergent property of systems of interdependencies, relations and interconnections and to isolate particular mechanisms will likely be transient or even impossible.

As argued above, social phenomena will tend to have a different mode of existence to non-social phenomena, and will be apprehendable in different ways. In particular, it seems impossible to arrive at knowledge of social phenomena in terms of isolation, either directly by constructing the closed system situations crucial for controlled experiments, or by directly perceiving the affordances of things. This is, in essence, the challenge faced by those with autism, that viewing things in isolation makes it impossible to understand much of the social world.

Thus, focusing upon those aspects of their relationship with artefacts that prove most challenging shows up the social, or the un-isolatable, aspects of technology. In effect, I am suggesting that the activity of those with autism acts a kind of index of the social, by drawing attention to those aspects of technology that are, at any moment in time, not isolatable and not understandable in terms of closed systems. And although those such as Pickering and others are surely correct to emphasise the highly complex and interconnected nature of different elements in any particular instance of technology use, we do not simply have to accept that there are no distinctions to be made. The activities of those with autism suggest a fairly clear distinction between those more mechanistic, intrinsic causal features or affordances with which they are relatively comfortable and capable of understanding and negotiating, and those less isolatable and relational features that they find so challenging.

Democratising Technology

To change focus again, I now wish to consider a second implication, this time concerned with the idea of democratising technology. More specifically, I want to consider briefly some of the reasons often given for pursuing some kind of democratisation project with respect to technology, and to provide further reasons for such a project that draw upon the arguments made above.

A central proposition here, a concern returned to in Chapter 11, is the idea that technology is thought not to be *neutral*. Roughly put, technology is neutral if it is simply some kind of efficient means to an end, taking on no particular character at all and independent of the uses to which some technology is put. Technology is effectively reduced to a value-free *means* serving different functional goals, where ends are purely subjective, coming from the users alone. This position finds many supporters, often amongst groupings that have little else in common, such as 'hard-nosed' engineers and the most relativistic of social constructivists.

This thesis of technological neutrality has been challenged in a variety of different ways, but a common feature of such challenges, touched upon in Chapter 5, is that objects are structured in particular ways that encourage or prescribe particular uses. Whilst this view is not deterministic, it does accept that different technologies provide

different kinds of constraints or inducements to action and that not all
conditions of action are the same. For example, Langdon Winner,
famously answered the question 'do artefact have politics?' in the
affirmative, by noting how underpasses were designed so as to only
allow access for small vehicles (not public transport which might
contain 'undesirables') to certain exclusive New York beaches
(Winner, 1980). Madelaine Akrich's formulation focuses upon 'scripts'
that can be understood as 'deposits' in an object left by the designer.
Designers inscribe objects with implicit manuals for use. Things thus, in
her words, 'co-shape the use that is made of them – they define relations
between people and distribute responsibilities between people and
things; technologies create frameworks for action' (Akrich, 1992).
Don Ihde suggests that objects have some form of technological inten-
tionality. Though his emphasis is less 'semiotic' than Akrich's, arguing
that it is not the signs that the object contains but the functioning of the
objects itself, the implications are very similar: objects are not neutral,
they co-shape the use that is being made of them, helping to colour the
practices within which technologies are realised (Ihde, 1990).
Elsewhere, as noted earlier, Feenberg makes similar points in terms of
what he terms technical codes. Here, design standards and norms of
construction are realised in the structure of objects in such a way as to
mediate the process whereby technology adapts to social change and so
conditions social life. Codes depend on a particular social context, but
then act as a mechanism by which prevailing values are literally materialised into particular objects (Feenberg, 2000).

There are clearly different emphases given to the analysis of such
problems, some emphasising capitalist hegemony and the struggles of
different groups for power and domination through the design of
technology, some presenting the situation as little more than accident.
But the upshot, argued perhaps most forcefully in the works of Winner
and Feenberg, is that the design of technical objects should not simply
be left to designers in the belief that important choices or decisions of
how technologies can be used will be made or adjusted at some later
stage. Instead, it is argued, democratic involvement is required early on
in the design stage. It is essentially this that the phrase 'the democratisation of technology' refers to. At least implicitly it challenges the
'common-sense' view that efficiency criteria are the only ones with
which technologists should concern themselves. Instead, the experience
and needs of everyday citizens should be channelled into the process of

design at an early stage. Often, recent environmental struggles and protests by particular constituencies, such as the disabled calling for greater access are used as positive examples of such democratisation processes in practice.

Rather than engage directly with this debate, my aim here is simply to add to the reasons for extending such involvement in design, drawing upon the discussion of autism above. To recap briefly, it is the social dimension of technology that those with autism have problems with. Or more specifically, those with autism seem to be drawn to those dimensions of technology most understandable in terms of closed systems, which renders much of the social dimension of technology challenging, unintelligible or possibly redundant. This attraction to technology would involve an attraction to (or relative comfort with) the workings of the intrinsic causal powers of things and devices where various social dimensions can most easily be ignored. However, there is likely to be a particular attraction to those aspects or moments in technology's development where isolatability and mechanical recombination feature most strongly. Thus there should be no surprise that those with autism are drawn to the design of technology, the isolative moment of technology, thus explaining the 'epidemics' of autism in those industrial and geographical clusters specialising in such design.

An implication of all this is that we seem to have additional reasons to worry about the adoption of a laissez-faire (or in Winner's terms, 'technologically somnambulant') attitude to technological design. A design group in which those with autism is overrepresented will inevitably issue in another particular form of bias, one where little attention is given to the (appropriateness of) ways that a device might be enrolled. More likely, the kind of enrolment envisaged may strike those without autism as impoverished, either socially, culturally or emotionally. Evidence for at least the more extreme versions of such a process do seem to be slowly permeating the 'collective consciousness'. Some examples might be the outspoken responses to Soylent, a liquid replacement for food intended to allow for meals to be consumed swiftly, shorn of their frequent use as settings for social interactions, and Google Glass, effectively wearing a webcam (for example see Hennessy, 2014).

Technological innovations that are relatively unresponsive to social needs will tend to reinforce and transform behaviour that is itself relatively unresponsive to social priorities. This may set certain groups against others, with those who feel most comfortable with undersocialised technology flourishing and those resisting both feeling, and being labelled, as conservatives or Luddites.

Thus there seem to be even more reasons for attempting to alleviate the design bias noted above, by involving different groups, and their experiences and needs, into the design process. The details of this involvement need not detain us here, though many of those noted above, especially Winner and Feenberg, provide a healthy stock of examples (for example see Feenberg, 2001, 2010, Winner, 1978, 1980). For present purposes, rather, the point is to illustrate that viewing technology through the lens of its relationship to those with autism not only challenges our understanding of technology but also our own relationship with technology.

Feminist Theories of Technology

Similar issues can be found if we consider feminist theories of technology. Whilst it would be difficult to give a comprehensive survey here of all the different feminist theories of technology, the more prominent theories do appear to share points of overlap or even convergence, which the above discussion has some bearing upon.[26] I shall give a very brief overview of some of these overlaps and point to implications of the discussion of autism given above where appropriate.

As noted in Chapter 2, the term 'technology' emerged in the way that it did, in part, because of the attempts of typically white, male engineers to hive-off and defend particular positions for themselves (Oldenziel, 1999). The resulting status afforded to engineers, simultaneously based upon a specialised education and ideas of manliness, served both to reduce the significance of 'everyday' technologies, such as corset or lace making, in which women excelled, as well as to redefine femininity as largely incompatible with technological pursuits (Cowan, 1976, Stanley, 1995). As well as highlighting such developments, various early feminist authors pointed to the importance of machines and weapons, cast in terms of the male activities of work and war, as the

---

[26] It is usual to distinguish at least liberal, socialist and postmodern feminisms as distinct perspectives. For excellent overviews see Wajcman (2010) and Layne et al. (2010).

paradigmatic examples of technology, suggesting that the 'mastery' of technology became viewed as a source of predominantly male power (Lohan and Faulkner, 2004).[27]

The contribution of these early feminist theories of technology was not only to highlight the association of men and machines but to suggest that such associations were the result of the historical and cultural construction of gender. However, in arguing that the main problem was that of men controlling neutral technology (liberal feminists) or power relations embedded in technology (radical feminists), there emerged a forceful (if often implicit) assertion that women's interests and needs are different from men's and, of course, rarely served by current technologies; by interpreting women as 'nurturing pacifists', a view of sex differences is reinforced that conflicts with ideas of historical and cultural specificity (Merchant, 1989). The same problems are present, if less obvious, with later accounts that focused upon the things that women had to give up (aspects of their femininity) in order to be accepted or succeed in technological (male-dominated) environments (Corneliussen, 2014).

More recently, feminist theories of technology have both challenged such ideas as a form of implicit essentialism as well as, for the most part, providing a more positive or optimistic account of technology. For example, although there is recognition of the dominance of men in information technology noted above, especially focusing upon examples of the macho and competitive environments of computer programmers and hackers (Turkle, 2005), this is set alongside a welcoming of the possibilities of information and communication technologies to empower women and transform gender relations (Green and Adam, 1998, Kirkup, 2000). Here the internet is picked out in particular as enabling an end to 'embodied bias' (Stewart Millar, 1998), and as blurring the boundaries between humans and machines as well as males and females (Plant, 1997). Although the contingency of woman's agency and need is made centre stage, there is at least an implicit proposition that the type of technology makes a difference, and that the increasing prominence of information technologies and biotechnologies relative to industrial technologies offers possibilities for liberating women if positively embraced (Haraway, 1985).

---

[27] Interestingly, although there is no space to pursue this here, similar arguments can be made with respect to science as well as technology (Harding, 1986).

In making such arguments, feminist theories of technology have been informed, and embraced, by the kinds of constructivist contributions as discussed above, especially in Chapters 1 and 5. The main emphasis is upon the ways in which technology and gender are mutually shaped and shaping, and in particular how each (technology and gender) are condition and consequence of the other (Wajcman, 2004, 2010).

There are some obvious similarities between feminist theories and the account set out above. One example is the central importance afforded to practice; for both feminist theories and the account I have defended, social relations and technology (especially the positionality of users of technology) are reproduced and transformed in and through human activities. Another example, one which is developed in the final chapter, is the way that technology is portrayed as embodying particular social relations, such as gender relations. Moreover there also seems broad agreement about the disruptive power of technology, whether or not this ultimately is beneficial or not for women (Johnson, 2010).

But there are also problems to be found within these theories of technology for which the above account may prove useful. For example, most feminist theories of technology have suffered from the problems noted above of constructivist approaches in general; in distancing themselves from any kind of technological determinism, feminist theories have ended up with a conception of technology that seems no different from any other social phenomenon and thus is of little use in explaining why technology might have any kind of special role in social change. Until recently at least, a stumbling block here may have been the importance of the material or physical in accounts of technology. However, much of the problem may have been removed in what might be termed a series of 'turns', such as the material, ontological and postconstructionist turns (see in particular Åsberg and Lykke, 2010, Lykke, 2010)).

Secondly, there seems to be a tension with respect to the way that essentialism is avoided in feminist theories. As noted above, in early feminist theories, there was an implicit acceptance that female 'characteristics' such as nurturing or pacifism, were being undermined by men, with the help of technology. Such ideas sit uncomfortably with ideas of historical and cultural contingency, at least of any thorough-going sort. This tension has been identified and criticised by more recent contributors, who have stressed the latter contingency as being the more important (thus rejecting any ideas of male or female traits or characteristics). Whilst this latter strategy seems largely correct, the account given in this chapter, at the very least, suggests some important ways in which feminist theories of technology should incorporate a more complex relationship of the relation between technology and gender.

Put most simply, if there is indeed a special relationship between those with autism and technology, and if autism is understood as a spectrum condition that, at the very least, has stronger effects on men than women, then there would seem to exist some kind of special relationship between most men and technology. Whilst I find it difficult to follow those such as Baron-Cohen in positing the existence of anything a like a 'male brain' and 'female brain', it does seem to be the case that biological differences exist, in terms of brain connectivity, etc., which are not equally distributed between the sexes and do have some bearing upon the kinds of relationships that are formed with technology. Very general observations that men, especially those interested in technology, tend to have greater preoccupation with details, tend to follow rules meticulously, etc., hardly seem far-fetched.

Such an association between men and technology is rather different from the feminist themes of power and control. But to suggest such an association is not to deny the importance of power and control, which clearly are as important to contemporary capitalism as they have ever been. Rather the point is to add an extra consideration in the mix, so to speak, in order to provide a more complex account. Recent contributors have been correct to emphasise that much of the difference between men and women in relation to technology has been positional and so very much historically and culturally contingent. Considering technology in relation to autistic characteristics introduces another dimension to the discussion, one in which historical and cultural contingency have less role to play.

Although there is little space to develop this argument here, it is illustrative to consider these ideas in relation to the democratisation of technology discussed in the previous section. Feminist theorists have for a long time focused upon the exclusion of women from design, as well as highlighted the lasting and constraining effects of such design on the lives of those excluded (Cockburn and Ormrod, 1993, Suchman, 2008). And more recently there have been a wealth of contributions suggesting how women may become more included in the design process (Bronet and Layne, 2010, Layne et al., 2010, Vostral and McDonough, 2010). The motivation for including women in the design process is typically presented as democratic or egalitarian; if only the playing field were level, more women would enter design and users as a group would be better represented. However, the argument of this chapter is that there exist other, complementary, reasons for encouraging their involvement. In particular, if there is a tendency for those with autism spectrum characteristics to be drawn to and to excel at design-based jobs (those most concerned with the isolative moment in technology's coming into being), then as argued above, there are reasons for suggesting that we will end up with the kinds of technology that we neither want nor need. Technology that genuinely responds to the wants and needs of all human beings and is appropriate to be enrolled or positioned in our lives in useful and meaningful ways will require that design is influenced at an early stage by as wide a group of concerns as possible. This idea, also, is returned to in the final chapter.

Concluding Remarks

This chapter has been concerned with the much documented but little explored idea that there is a special relationship between technology and those with autism. In particular, it has been concerned with the idea that this special relationship might tell us something about technology. The chapter has drawn upon current accounts of autism and in particular on that theory of autism which seems best to accommodate both what we know about the condition and the main other existing theories of autism. If a central feature of autism is an inability to deal with the predominantly 'open' nature of social interaction, it is easy to see why, for those with autism, some aspects of technology are attractive and some are not. More specifically, it is the initial moment of technical activity, most clearly concerned with the isolation of intrinsic causal powers of things, the on-going harnessing of such causal powers, and the more mechanistic forms of recombination of different mechanisms and devices, which are the aspects of technology that those with autism are likely to find more attractive and to excel at.

Formulating things in this way has implications that range from the analytical to the political, as the examples given above demonstrate. As yet, there is little, or no, significant technology research that considers the activities of those with autism. Additionally, for those studying autism, the question of any kind of relationship with technology is left very undeveloped. For the most part, the 'excellent relations' to physical objects, observed by Kanner is accepted pretty much at face value and the conception of technology used rarely goes beyond a simplistic idea of material devices. It is hoped that this chapter, at the very least, will stimulate more consideration of what I believe to be a fascinating set of connections between, as well as a lens with which to view, both technology and autism.

10 | *Technology, Recombination and Speed*

In this chapter I want to develop the idea, introduced in Chapter 6, that recombination plays an important role in explaining some of the special characteristics or properties that are often associated with technology. That recombination plays an important role in our understanding of technology is a suggestion that has gained prominence in recent years within some strands of mainstream economics.[1] However it is not a new idea, and can be traced back at least to contributions to the sociology of invention literature of the 1920s and 30s. Moreover, as noted above, a focus on recombination was also central to the work of Clarence Ayres. The aim of this chapter is to review the different conceptions of recombination that emerge in these different literatures and highlight their strengths and weaknesses. I argue that some conception of recombination marks a positive development in all these traditions. However, I also argue, drawing upon ideas from earlier chapters, that for a conception of recombination to be of much use it must both be given a more ontological formulation and be combined with a more complex conception of technology. The advantages of this reformulation are illustrated by briefly considering the contention that modern societies can be understood to be speeding up or accelerating.

From Production Functions to Recombination

As noted above, a common criticism of standard mainstream economics is that it pays little attention to the study of technology, or more precisely there is little concern with the nature of technology. Within mainstream accounts, technology's presence is felt via the stipulation of different relationships between inputs and outputs. At most, technology is conceptualised as some kind of 'menu' that is intended to capture

---

[1] And perhaps surprisingly, via some of these mainstream contributors, into current philosophy of technology accounts (for example see Vaccari, 2013).

constraints placed on human action in relation to production (Metcalfe, 2010). These constraints are left unexplained, in the sense that their explanation is thought to lie outside the purview of economics proper. Given these constraints, however, economists are concerned with what can be said about the relations of inputs and outputs, information typically summarised in terms of the shapes and properties of production functions. In this case, questions about technology reduce to questions about the efficiency with which inputs generate outputs, the relative proportions in which the inputs are employed and the ease of substituting one input for another. None of these issues requires much consideration of what technology might be.

The story is only a little more complicated if we turn away from the more standard, textbook treatments of mainstream economics. The effective banishment of technology from economics 'proper' became particularly embarrassing with an acceptance amongst (mainstream) economists that certain evidence – historical (Abramovitz, 1989, 1956), and econometric (especially Solow, 1957) – suggested that technology was the main contributor to economic growth.[2] Thus the prime generator of one of the most important economic variables was something about which economists had little to say. In attempting to reclaim some of this subject matter for economics, attempts were made to ground the growth-inducing characteristics of technology in the peculiar nature of knowledge.[3]

The most prominent example here is provided by the work of Kenneth Arrow, who's approach was to characterise knowledge as unlike other factors or inputs to the production process (Arrow, 1962, 1969).[4] After acknowledging the problems with this response, there

---

[2] And indeed that much could not be explained in terms of the usual framework of shifting production functions.

[3] Although a range of mainstream accounts have also located technology's special features in knowledge, it is not always clear whether there is ultimately more to technology than knowledge. Although most do not consider the matter directly, for some (for example for Mokyr (2000, 2002)) technology simply is knowledge.

[4] In particular, knowledge was thought to be non-exhaustible, cumulative and not easily appropriated. Technology, understood as in some way dependent on knowledge, takes on the same characteristics such that it is not subject to decay or becoming used up with use (successive generations of agents continue to take advantage of previous technologies). Appropriability becomes a part of this story as agents have an incentive to keep adding because they can appropriate only a small part of the benefits that are generated from new knowledge. Although relatively ingenious as a way of saving the main features of the mainstream

was then a general move to add different forms of complexity to the standard model to account for the now well-documented differences in knowledge creation, economic growth, path dependency, etc.

It is in this context, as an attempt to add such complexity, that recent contributions focusing upon recombination are best understood. In particular, various writers have tried to locate the peculiar properties of technology in the fact that it seems to evolve by some process of recombination. The essential idea, in the mainstream formulation at least, is very simple: whatever technology might be, there is a common or pervasive dynamic to the way that it develops that results from a more or less continual recombination of the elements that constitute it. A prominently cited example is the work of Martin Weitzman (1996, 1998). Technology is understood to be constituted by components or modules which, when combined, are then ready for further (re)combination. The crucial part of the argument is that the more components there are, the more possibilities there are for recombination. Indeed, the possibilities for recombination, and so technical innovation, are thought to develop exponentially. For Weitzman, this 'recombinant growth' accounts for the dramatic increase in economic growth witnessed after the Industrial Revolution, once the recombination of technological components started to take place 'in earnest' (Weitzman, 1996).

According to Weitzman, there is no need to clarify exactly what it is that is being combined. In fact, there is little concern at all with actual processes by which new technologies might come into being. The mainstream orientation, which effectively brackets any consideration of the nature of technology, is evident here. One exception to this orientation, however, is the work of Brian Arthur, whose account is far more sophisticated and well worth considering in some detail.

Arthur and Combinatorial Evolution

Recombination is a central part of Arthur's recent contributions on technology. And technology, Arthur suggests, is central to our accounts of the world, in the sense that it is our technology that distinguishes

model, the limitations, even on its own terms, are well known. For example, it is difficult to explain variations in different rates of growth, rather than homogeneous and steady rates of improvement of technology across agents, regions and times. For a detailed account see for example Antonelli (2008).

humans from other forms of life and has enabled modern societies to be so different from previous societies, such as the Stone Age or the Middle Ages (Arthur, 2009, p. 10). The key to understanding how this is so, Arthur suggests, lies in understanding technology dynamically, as being generated and constituted by processes of recombination.

There are several aspects to this idea. First, novelty rarely arises from some isolated inventor. Rather it tends to involve rather mundane processes of piecing together existing components and ideas. Secondly, the stock of existing technology provides the components from which future recombinations take place; thus, the progress of technology is in some sense cumulative. Arthur terms this process of development 'combinatorial evolution', involving a continual 'bootstrapping from the few to the many and the simple to the complex' (ibid., p. 21). Such evolution can be contrasted with Darwinian ideas of evolution because in nature fresh new forms do not appear abruptly (as jet engines or radar have done), and there is no mechanism for new species to adopt the characteristics of a wide range of existing species (the wings from a bird, the head of a cat, etc.,). In other words, there is more to technology than variation and selection.

Arthur uses these basic ideas to develop an account of how technology evolves, how it changes over time, how novel technologies come about and how novel innovations are distinguished from 'standard engineering'.[5] His account is thoughtful and benefits from a clear enthusiasm for technology, its history and how particular instantiations of technology actually work. In terms of developing an account of recombination, its main advantage is probably the wealth of examples he provides to illustrate exactly how these recombination ideas have worked out in the past. However, it is not clear how specific Arthur's conception of combination is to technology. Can non-technological things be combined in the manner Arthur discusses? Is there something about this thing called technology that facilitates recombination? Without addressing such questions, it is not clear that technology can

---

[5] He also uses it to construct a view of the economy not as a container of technology but as something formed from technology, with technology providing something like the skeletal structure from which the economy develops. Central here is the idea that technology provides the needs, requirements and problems that the economy has to solve. Thus, via a process of overcoming the immediate weakest links (Hughesian reverse salients) and Schumpeterian disruption, some kind of co-evolution is established between the economy and technology.

be distinguished from other social phenomena, so it is unclear how technology comes to play the very special role Arthur wants it to in his account.

At times it seems that technology's special properties depend upon what is being combined. Critics argue that Arthur only considers material devices (e.g. Dosi and Grazzi, 2010, Antonelli, 2008). Certainly his favourite examples are such things as jet engines, radar and electricity generation. Moreover, underlying these processes of recombination, Arthur suggests, there must always be some physical or natural world mechanism (such as gravity or the boiling point of water – he terms these 'phenomena') that lies at the heart of the technology in question. And thus, these recombinations are really recombinations of devices or mechanisms that harness such phenomena. In this sense technology can always be understood as 'nature organised for our purposes' (ibid., 213).[6]

However, on viewing Arthur's position as a whole, things are not so clear. Most often, Arthur returns to his definition of technology as anything that serves as a 'means to a purpose', a definition that actually rules out very little. Arthur accepts this and suggests that business organisations, legal systems, monetary systems, contracts and even symphonies are all technology because they can all be understood as a means to some purpose or other (ibid., p. 54). He goes on to add that these may not *feel* like technologies because the phenomena being drawn upon in all these examples are not physical but organisational or behavioural, such as the trust various parties have that contracts will be adhered to, promises kept and value maintained. Arthur considers other phenomena, such as methods or algorithms, which are based upon logical phenomena, rather than organisational or behavioural ones. But Arthur makes little real argument for including these other

---

[6] When defining technology more precisely, Arthur actually insists that there are three quite different uses of the term. Here he distinguishes technology as 'a means to a purpose', as an 'assemblage of practices and components' and as the 'entire collection of devices and practices available to culture' (ibid., 29). Essentially the distinction is between individual devices or methods that do something in particular (e.g. detect objects, power aircraft) and bodies of devices and practices (such as biotechnology or electronics) which are able to do all manner of things, but in particular manage to focus on combining individual 'means to a purpose' in different and useful ways. But in each case, what is essential to technology is that phenomena are captured or put to use via this process of recombination.

phenomena or not, simply stating that: 'we should remember that these too are technologies if we choose to see them this way.'

Although Arthur does not give much attention to demarcating the technological, he does offer an implicit argument when he considers the role of internal replacement and structural deepening in the tendency of technology to become more complex. The drive for technological change, he suggests, often derives from the need to address the weak links in the technology itself. These lead to replacement of particular components or the adding-on of compensating modules or subsystems to 'workaround' the poorly functioning component. The only cost Arthur considers, for technology proper, is the financial costs of development. However, for those technologies that rely on or harness organisational or behavioural phenomena, the more important cost is that these growing 'improvements' may be at the expense of complication and bureaucracy that may be 'inefficient' in some sense and difficult to get rid of (ibid., p. 138). In other words, the growing complexity itself produces costs in social systems that are not present in physical devices.[7] Although I believe this idea hints at some important refinements to Arthur's position, unfortunately, this point amounts to more than a throwaway, and Arthur seems generally uninterested in the implications of such differences. But because of this, he is not well placed to answer the initial questions he asked himself. In particular, what exactly accounts for the distinctive position technology assumes in modern societies and how does it account for the momentous societal changes witnessed? If, as Arthur argues, all kinds of things can be recombined, why has development taken place in the manner that it has? There is a (familiar) tension here: technology assumes a central role in Arthur's story, but it seems indistinguishable from other (social) phenomena.

In order to pursue this question more successfully, I believe we must shift our attention to the sources of Arthur's ideas, the sociology of invention literature of the 1920s and 30s, where a concern with such issues was both explicit and central.

---

[7] Although I may be reading too much into Arthur's comments, he seems to suggest that at the centre of the issue is the relative atomistic or organic nature of the system in which these changes take place. The more atomistic it is, the easier it is to add and replace components; the more organic it is, the more complicated, messy and costly this process is likely to become.

From the Sociology of Invention to Technical Artefacts

The best-known figures of the sociology of invention literature are William Ogburn (1922, 1933, 1938) and Collum Gilfillan (1935a, b).[8] These authors were primarily motivated by an interest in the factors causing change in social and cultural institutions. Contributors were primarily sociologists who were impressed with the effects that technological change was having on all kinds of social phenomena but also worried that social theory, and sociology in particular, had essentially ignored such changes. As noted in Chapter 2, the term 'technology' was itself undergoing dramatic changes in use in the early twentieth century, and tended to be applied vaguely and inconsistently. Thus the terms 'invention' or 'material culture' were often used. However, it does seem that the referent of these terms is what today we would call technology (Godin, 2010).

Although Arthur presents his ideas as 'working from scratch', they are strikingly similar to those popularised in the 1920s and 30s in the sociology of invention literature. Indeed, whilst Arthur dismisses the sociology of invention as being undeveloped, there is very little in his formulation of recombination that does not appear in this earlier literature. For example, whereas something like an evolutionary story is being told, there is significant effort given to contrast social change with biological change (in particular see Ogburn, 1922). The role of the individual inventor, having 'eureka' moments in some lonely laboratory, is similarly downplayed in favour of a conception in which invention takes place over long periods of time with a succession of different contributors adding exponentially to existing knowledge. Central, again, is the idea that invention involves a more or less constant process of recombination of what currently exists, selectively recombining things in different ways. An even greater wealth of examples is provided of the manner in which different inventions came into being, consistently applying the ideas of recombination to explain the patterns of events observed. A particular focus is given to two sets of examples, one (predating Weitzman's work) focusing upon the idea that the numbers of inventions increases exponentially and the other focusing upon the observation that the same invention tends to be made at the same time by different inventors in different places. Both

---

[8] Though similar ideas can also be found amongst the work of anthropologists such as Kroeber (1948) and Chapin (1928).

observations are exhaustively and comprehensively argued to follow from the basic idea that it is the availability of potential components ready for recombination that drives new inventions (especially see Gilfillan, 1935a).

As already noted, however, there was a different motivation for the sociology of invention contributors, namely, to understand the nature of institutional and cultural change. And this motivation required a more concerted attempt to distinguish between different kinds of recombination. Crucial to this was a profound interest, especially as illustrated in Ogburn's work, in the manner in which societies adopt (or fail to adopt) and react to the growing number of inventions. His well-known concept of a cultural lag is essentially an attempt to accommodate all the reasons why society is often slow to react to such changes. He saw this very much as a new problem (of his day), largely because in earlier times inventions came about so slowly that they were relatively easy to keep up with. But in modern times, he argued, the speed or manner in which new inventions are kept up with becomes a matter of some significance. Much, for Ogburn, depends on attitudes to 'newness' and large parts of his discussion deal with this, along with discussions of the general fear of the unknown (Ogburn, 1936a).

However, despite Ogburn's concern to distinguish the social and institutional in opposition to the technological (and account for the gap between them), there is little or no discussion of the different kinds of things that inventions might be or why they might be subject to some kind of continuous impulse to change, whereas the institutions he focused upon were not. Whilst he gives many important examples, and draws extensively on Gilfillan's exhaustive descriptions, there is little general discussion of the kinds of things that are 'combined and accumulated'. Thus despite both initiating a more ontologically differentiated approach, distinguishing social and cultural institutions that either do or do not adapt to new inventions, and generating a very fruitful and influential approach to the study of invention (or innovation), there is still no real reason given to explain exactly why material culture should have different dynamics or properties than non-material culture. In short, the sociology of invention only takes us a little beyond Arthur's 'means to a purpose'. This absence however, as noted in Chapter 5, is just the issue addressed in the work of Clarence Ayres. Indeed, as far as I know, Ayres, a good 70 years before Arthur, was the first economist to explicitly introduce the ideas of the sociology of invention into economics literature.[9]

From Tool-Use to Instrumentalisation

I argued above that the strength of Ayres's account lies in his attempt to provide an ontological conception of technology based on the distinction between tools and icons. The focus in this chapter is on how this distinction matters for the idea of recombination. It should already be clear that Ayres employs many of the arguments to be found in the sociology of invention literature, which also predate most of Arthur's work.[10] However, Ayres's distinction between tools and icons allows him to go much further than Arthur, Ogburn or Gilfillan in explaining how and why recombination plays the role that it does, and in explaining the distinction between technology and other kinds of phenomena.

In brief, Ayres elaborates an ontology of artefacts in which tools gain some kind of transcultural characteristics because they are not only materially and physically durable but because their causal powers are essentially isolatable, thus making the tools themselves largely separable from, and useable outside of, any particular context of use. Icons and fetishes have properties too, and can be harnessed in a variety of different ways. But the causal powers of such material objects are not isolatable from the communities and the networks of agreements and meanings within which they have meaning and potency. These distinctions were accommodated within the conception of technology developed above in particular by adding the moment of recombination to that of isolation and positioning that were the central concerns of Chapters 4 and 5 respectively.

---

[9] Although it clear that Ogburn's work was well known to many of the American institutionalists, especially those writing most between the wars (Rutherford, 2011).

[10] As noted in Chapter 6, Ayres pays particular attention to the idea that there is an exponential growth in inventions that is not easy to locate in improvements in human capabilities, such as intelligence, knowledge, increased skill. Rather, the explanation for such exponential growth lies in the idea that all inventions are combinations of previously existing devices; what appear as inventions are usually the end product of a long series of combinations. Such inventions are not due to the 'magnitude of the soul of the "Gifted ones"' (1978 [1944], p. 115). Ayres repeats that these are most often little more than the rather mechanical combination of existing devices, materials, instruments and techniques (1978 [1944], p. 113).

On this account, recombinability in the sense employed by Arthur, Ayres and the sociologists of invention, can be seen as primarily concerned with the more atomistic aspect of the process whereby isolated properties or mechanisms are recombined to produce functional devices (roughly what Feenberg terms systematisation). The problem with focusing so heavily on recombination, understood in this way, is that it effectively ignores processes of isolation and positioning. Positioning working devices into networks of use, or more generally the lifeworld, something which is essential for technological artefacts to be realised as technology, is not part of the story in recombination accounts, even the more ontologically sophisticated version given by Ayres. Moreover, as the idea of isolation is also missing, there is no attention given to the fact that the process of recombination might involve filtering out particular characteristics of the components used in any recombination, as highlighted by Feenberg. In other words some things might not survive the recombination process.

Incorporating this conception of recombination within the account of technology developed above, however, seems to have a series of advantages. First it is possible to ground or provide an ontological account of different tendencies in the generation and adoption of new technologies. In other words, it is possible to make a clear distinction between processes that, on the one hand, tend to produce more devices or technological artefacts and, on the other hand, lead different technological artefacts to be realised as actual functioning technology. On the account suggested here, a tendency to proliferation exists simply because, as recombination theorists suggest, there really are stocks of isolatable, endurable objects that can be recombined to generate new devices. The idea that there is an exponential increase in the number of possible recombinations, and that this tends to create conditions in which new technological artefacts can come into being more rapidly, makes a good deal of sense in this context. However, it is only part of the story, given that enrolment takes time and is subject to all kinds of constraints.

To argue that enrolment takes time, however, is not simply to return to the ideas of cultural lag and ceremonialism of Ogburn and Ayres. It is not simply that 'resistance' to technology is something that is problematic and to be overcome. For one thing, adopting a new technology may not be the obviously best thing to do. The intrinsic filtering that accompanies the isolative moment may involve the loss of elements that are not wanted, not needed or may be harmful, in which case it may be best not to try to enrol that new technology. For whatever reason, the processes by which enrolments take place require time and deliberation. Although this point is implicit in Ayres's own work, he was too keen to give a negative gloss to 'the ceremonial' to take his insights very far. If the social world is constituted by community agreements, acknowledgements and meanings, all of which take time to become established and are reproduced only through the more fragile processes of daily reproduction and transformation, then enrolment is unlikely to be something that will happen quickly in a way that does justice to the meaningful and knowledgeable ways that people tend to enrol objects if given significant periods of time to do so. Capable and meaningful positioning or enrolment will incorporate values about all kinds of things from the importance of mealtime gatherings, to the effect of computer games on the development of social skills, etc. As will be focused upon in the following chapter, there are choices to be made about how (and indeed whether) to enrol a variety of different technologies. To suggest that such things take time is not to commit to the idea that all failure to adopt new technologies is necessarily problematic.

Technology and Speed

I now want to illustrate these ideas by considering the contention that modern societies can be characterised as in some sense speeding up.[11] Many different accounts draw attention to the ways in which such speeding up is thought to be experienced or detected. A wide range of examples is provided by Gleick (1999). Work, eating, the length of parliamentary speeches and adverts, the time between adverts, the time between moving house or job or changing car, the time spent out walking, concentrating on any one thing, sleeping, playing, talking with family, engaged in any leisure activity, indeed spent on just about everything, Gleick suggests, is speeding up. His findings seem well corroborated by a range of similar accounts (Rifkin, 1987, Robinson and Godbey, 1999) and will seem familiar and plausible to most of us.

---

[11] The phenomenon of speeding up is often termed 'social acceleration' (Rosa, 2013). Whether there is an important distinction to be made between speeding up or accelerating is not one I shall pursue here.

The feeling that much is speeding up is, of course, not a new experience, and seems to have been a feature of modern societies, at least since the middle of the eighteenth century (Koselleck, 2004). But until recently, the subject of speed, or time more generally, appears to have been largely absent from social theory (Adam, 1990). Recent accounts, however, are starting to focus on a range of related questions (Rosa and Scheuerman, 2009, Rosa, 2013). For example, what is the relationship between the speeding up of so many daily activities, as noted by Gleick and others, and the perception or experience of a lack or scarcity of time?

Many of these accounts are not so much concerned with increasingly fleeting events, or a scarcity of time as such, but with a range of phenomena that speeding up appears to bring with it. One example is the emergence of new kinds of activities, such as 'speed-dating' and 'walk-through funerals'. There is also a concern with the implications of more structural changes, viewed as constituents of societies rather than events taking place within society. For example, there is concern with the way that various societal or structural 'givens' become increasingly unstable or ephemeral. Attitudes, values, lifestyles, all manner of social relations and obligations, characteristics of groups, classes, languages and institutions such as the family and work are all perceived to be speeding up or changing at increasing rates (Rosa, 2009, see also Appadurai, 2001).

Such changes are, thus, not only thought to be speeding up, but speeding up to such an extent that they are producing qualitatively different lifeworld experiences to those of previous generations. For example, it has been suggested that before modernity these basic institutions remained constant between generations. In early modernism, they change from one generation to the next. In late modernism, however, they change for each generation, disrupting, destabilising and making unreliable a range of different aspects of society that would otherwise provide the basis of effective and meaningful decision-making (see especially Rosa, 2009). Expressed differently, the present becomes disconnected from the past, this 'contraction of the present' making a poor basis for forming expectations of future actions (Lubbe, 1994, 2009).

In attempting to explain such processes of speeding up, there is much agreement that at least part of the driving force is the nature of modern capitalism. In this context, it is hard to ignore the explicitly Marxist contributions that emphasise the continuing need for capitalist producers to realise profits ever more quickly, reduce the time required to produce a commodity, accelerate the time between new innovations (but increase the time over which monopoly profits are available), raise the amount produced from each unit of labour power bought, rapidly transport goods around the world ever more quickly, etc. Of course, such ideas are not new (see for example Simmel, 1991, 2004) but they have recently been developed in novel and interesting ways, for example in terms of the accelerated experience of time and the reduced significance (or annihilation) of space, or time-space compression (Harvey, 1989) or the temporal dimension of globalisation (Jessop, 2009).[12]

Despite the complexity of the explanations of social acceleration that have emerged, however, the role of technology remains generally undeveloped. At the very least, it can be said that there is a clear ambivalence with which technology is treated in 'acceleration' accounts.[13] For example, it is common to view technology as the ultimate source of society's speeding up, whilst at the same time refusing to couch general discussions in technological terms.[14] Rosa attempts to explain a lack of general interest in technology by returning to the idea that technology should in principle save time.[15] The fact that we experience less time, Rosa argues, means that increased technological change must be an

[12] Although it is fair to say that most explanatory accounts seem to at least give a prominent role for the increased need to produce, circulate and consume more goods associated with modern capitalism, other explanations exist also such as those focusing upon cultural changes. A prominent example here is the rise of secularisation. If there is little belief in any form of afterlife, then the goal becomes simply to fill 'this' life with as much as possible. The good life becomes the full life (Rosa, 2009).

[13] I should stress that in using the term acceleration accounts here, I do not want to suggest any similarity between the ideas under consideration and those of 'accelerationist' accounts, which typically refer to something rather different – such as, that we should embrace the worst excesses of capitalist development as quickly and wholeheartedly as possible so as to speed up the process of 'coming out the other side' of capitalism; see for example (Noys, 2014).

[14] One possible counter example is the concept of 'dromology' created by Paul Virilio. In his account, military technology tends to encourage the process whereby decision-making concentrates more and more in the hands of 'experts' that may be relatively unresponsive to the desires and needs of most citizens (Virilio, 2006).

[15] But this is not always the case. The obvious example is Albert Borgmann's account of growing computer use. As processing speeds and digitalisation increase, using a computer to do a range of tasks more quickly holds out the promise of freeing us all up to be able to go and do something more worthwhile,

effect of speeding up, not a cause (Rosa, 2009). Elsewhere the argument is made that attempts to speed up such aspects as communication and technology occurred before modernity (Koselleck, 1985). In effect, because there is nothing new in the technological advances occurring, the focus should be upon the manner in which such technological changes are received, and so should be upon social, cultural or institutional changes.

Perhaps the most dominant reason for the reluctance to make technology a focal part of the analysis of speed or acceleration is, however, a reaction to the excesses of technological determinism and in particular the simple preoccupation with 'progress' thought to be heralded by particular levels of technology. Yet, as earlier chapters indicate, neither determinism nor an overly optimistic conception of progress need be invoked to provide a more central role for technology in explaining the nature of society's speeding up. Several points are worth making here. First, it is easy to explain a tendency towards the increased proliferation of new technical devices in terms of the exponentially increasing possibilities for recombination. Importantly, of course, this need not be seen as the ultimate cause of such changes, but one mechanism amongst many. Thus there is no tension between the idea that recombination provides a tendency towards the proliferation of new devices and the claim that institutional and structural conditions of capitalism are such that there is a continual tendency towards reducing production times, convincing people that they really do require ever more products, devices, etc.

Ultimately, however, there are pressures not only towards the increased proliferation of new devices, but also for them to be realised via some kind of positioning or enrolment. Not only does modern capitalism depend on a growing turnover of new devices, gadgets, etc., to make its profits, the promise of new opportunities and abilities being opened up with the enrolment of such devices constantly serves to encourage the more or less continual enrolment of new devices (Borgmann, 1984).[16]

However strong these pressures are, there is no obvious reason for believing that there will be an increase in the ability of users of

---

such as take a long walk or paint a picture. The reality, however, is that all it tends to free us up for is more time on the computer (Borgmann, 1984).

[16] See also the following chapter.

technology to enrol new technical objects in meaningful or satisfying ways. Such processes, as noted by Ogburn, Ayres and Feenberg, in their different ways, take time. In this case, it seems easy to account for the growing perception that devices or objects keep arriving on the scene at a speed that makes meaningful, satisfying or fruitful enrolment difficult to keep up with. Given the benefits to capitalist producers of encouraging a growing use of such objects, and the 'promise' of new devices highlighted by Borgmann and others, it is unlikely that new technologies will fail to be taken up (enrolled or positioned in some way). Rather, it seems more likely that processes of enrolment will become increasingly superficial and transitory, following some minimum requirements for particular objects or devices to be made directly useable. It is also likely that such superficial enrolment will tend to become the norm.

Thus technology's tendency to proliferation is likely to be experienced as a speeding up of social life, which is often accompanied by a sense of the increasing trivialisation or superficiality of the lifeworld. Such concerns are a long way from the narrowly defined focus upon economic growth of mainstream economists or the negative effects of institutional inertia of Ayres. But they are the kinds of concerns that spring to the fore once a concept of recombination is combined with the account of technology developed above, in particular with ideas of isolation and positioning. Moreover, these concerns seem to be well-known and heavily researched themes about which, as yet, the ideas of recombination or instrumentalisation have so far had nothing to say.

Concluding Remarks

Recombination appears to be a very significant part of the story of how technology comes into being and assumes the role in modern societies that it does. As Ayres, Arthur and the sociology of invention contributors illustrate, even when the focus is on very discrete, apparently radical inventions, these turn out to be the results of a series of incremental changes involving long processes of more or less continuous and often very mechanical processes of recombination.

This chapter has argued that, although important, recombination is only part of the story. More specifically, I have argued the merits of adding a more ontological account of recombination with the conception of technology developed in earlier chapters. Ayres's account of recombination, based on a distinction between tools and icons (which itself relies on an implicit conception of the relative isolatability of the components being combined) has been shown to be quite compatible with (and in need of) a more complex account of technology, one that explicitly acknowledges different moments in the process in which technology comes into being. In terms of this wider framework of ideas, recombination only seems relevant to one aspect of the process, this being the more atomistic form of secondary instrumentalisation that is set in play once mechanisms, objects and devices have been separated off and before the resulting artefacts, devices, etc., are positioned or enrolled in social, cultural and institutional systems. Recombination relates to and develops our understanding of this step. However, it is an important step in so far it has proved very difficult to explain the proliferation of technology without resorting to unwarranted determinism.

Moreover, once a wider focus is adopted, the emphasis moves away from the preoccupations with negative effects of institutions on technology as seen in the work of Ayres or the preoccupation with growth that can be seen in mainstream economics. The issues that seem to require attention result, rather, from the dynamic relationship between the out-of-sync nature of the processes of isolation, recombination and positioning or enrolment. Only one example has been focused upon here, namely, the role that technology plays in explaining the tendency towards the speeding up of modern societies. I have argued that whilst it is possible to explain the proliferation of new technical objects in terms of recombination ideas, there is no particular reason why our ability to enrol and absorb such devices in meaningful or socially beneficial ways will also speed up or improve in any sense. Thus a major issue raised by the idea of recombination, but not addressed by recombination theorists, is how to respond to the potential 'out of phase-ness' of social and cultural institutions with new technological developments. This is not quite the cultural lag of those such as Ogburn, nor the ceremonialism of Ayres, though there seems potential to develop these ideas in a complementary manner. And to the extent that such accounts can be used to explain continued economic growth, it may be that the explanation has as much to do with the way that particular social and cultural institutions are undermined as it has to do with the recombination of devices as such.

Combining a more ontological version of Ayres's distinctions within a more general account of technology along the lines argued for above seems to provide the beginnings of an important contribution, but it also raises other issues. As a great deal of the literature referred to in this chapter suggests, there is a general acceptance that technology is crucially important for explaining a range of dramatic social, cultural and institutional changes. However, given the flight from determinism, there is a reluctance to spend much time considering any general features that technology might have. In the next chapter this issue is addressed by considering two very different accounts of technology, both of which focus upon the general implications of technology acquiring a greater and more dominant role in modern societies.

11 | Marx, Heidegger and Technological Neutrality

To return to themes of the first chapter, perhaps our most common or most shared experience of technology is as an agent of change, that is, as an external prod to our normal or routine ways of doing things. Indeed, it was largely in order to capture this role that the term technology became a *keyword* in social theory. Moreover, it is perhaps understandable that so many accounts of technology, at least until recently, have attempted to analyse or explain this experience in terms of the apparent inevitability of such changes and as part of some more general evaluation (moral, ethical, etc.,) of the implications of the actual changes technology introduces. The main motivation of this chapter is the belief that although these issues are as relevant today as they have ever been, they have recently fallen out of favour largely because of the way they have been formulated in earlier accounts. Specifically, various attempts to address these issues have been interpreted, often understandably, as a form of technological determinism. Given the implausibility of any form of determinism, including its technological variant, accounts that might be labelled as such have tended to be dismissed or ignored.

Linked to this dismissal of technological determinism has been an implicit acceptance of some kind of technological neutrality or instrumentalism. Here, technology is viewed as neutral or instrumental in the sense that any implications that follow from the introduction of some particular technology take the form that they do simply because of the particular context or manner in which they are applied. There is, then, nothing of interest that can be said about the properties or characteristics of technology itself, or in general.

In this chapter, I argue that in dismissing accounts typically labelled as technological determinist and at least implicitly embracing a form of technological neutrality, recent contributions have lost sight of some of the important questions concerning technology, as well as some of the conceptual resources that might be used to answer such questions. At the very least, the aim of this chapter is to put such issues back on the agenda, although in rather different terms to those usually employed. To this end, I focus primarily on the work of two significant contributors to the study of technology whose work tends to be portrayed (and, more recently, readily dismissed) as some form of technological determinism, namely that of Karl Marx and Martin Heidegger. It is worth noting that although the writings of both authors are currently very unfashionable, they are also regularly returned to. Explaining this continuing, if seemingly reluctant, appeal is something else that I hope contribute to. However, I shall make no attempt to provide an exhaustive account, let alone defence, of the work of either Marx or Heidegger. Rather the aim is to recast, somewhat selectively, some of their contributions on technology. Although the result does highlight some kind of complementarity between the ideas of Marx and Heidegger, the kind of complementarity I identify is quite different from the kind usually found in the literature.

Where Heidegger and Marx are considered together, their work tends to be viewed as alternative explanations of the 'ills of modernity', or the problems of capitalism. One of the more sympathetic accounts of such differences is provided by Collier (2003, Chapter 7). Collier, although finding much of value in both Heidegger's philosophical contribution and his 'worry' about modernity's basic problems, argues that Heidegger is wrong to lay the blame for these problems at the feet of technology. Instead, it is the 'exchange-value driven-ness' of capitalism that produces the worldview in which everything is viewed as a stockpile of resources ready to be utilised; technology itself is not the problem.

Where Marx and Heidegger have been considered in a more complementary manner, the focus has again been upon the problems of capitalism. Perhaps the most prominent examples of this approach are to be found in critical theory. Starting from Lukács, there has been a focus upon the concept of reification to explain not only how capitalism survives, but does so by representing itself as the only possible organising system for human societies (Lukács, 1971). Critical theory is most often viewed as a combination of Marx's conception of fetishism and Weber's conception of rationalisation. However, the phenomenological influence of Heidegger is also clearly present. This is perhaps most obvious in the work of Marcuse (a student of Heidegger), who explicitly incorporated ideas from both Marx and Heidegger and extended the ideas of reification to media propaganda, consumerism, etc., in which effectively all of social life had become commodified.

Although important background, I am not considering the work of Marx and Heidegger in either of these ways. Rather I restrict myself solely to a consideration of their contributions to a theory of technology and relate these to the account set out above. In so doing, I want to marshal not only the arguments of each author against a conception of technological neutrality but also show other ways in which my account of technology might usefully be drawn upon and extended.

One further qualification needs to be made from the start. It should be clear from earlier chapters that I do not want to defend any form of technological determinism. As a general position, if humans have anything worthy of the name choice (and I believe that they do) then technological determinism, as a thesis about the world, must be incorrect. My interest in the theme of technological determinism is that historically it has come to 'house' many of the arguments I am interested in, as well as suggesting why such arguments have been dismissed. In particular, those aspects of Marx and Heidegger's account most usually held up as examples of technological determinism, although not without their problems, are I believe far too interesting and valuable to be dismissed and forgotten.

Before directly addressing the issue of technological neutrality, I shall first review some of the main features of each author's contributions to a theory of technology.

Marx, Machines and Domination

Those commentating on Marx's writings on technology have concentrated upon two commonly recognised tensions or ambivalences. First, Marx is often portrayed as a technological determinist but also as an antideterminist (Rosenberg, 1976). Secondly, Marx is also often portrayed as having a view of technology that is both dystopian and yet fails to shake off an Enlightenment enthusiasm (Kirkpatrick, 2008). Let me elaborate.

Marx's work is perhaps the most commonly cited example of a form of technological determinism (Hansen, 1921, Heilbroner, 1967, see also Lawson, 2007). His famous statement that 'The hand-mill gives you society with the feudal lord; the steam-mill, society with the industrial capitalist' seems to be as clear a statement of technological determinism as anyone might want (Marx, 1955 [1900]). The case is further strengthened by Marx's discussion of productive forces, the following being the best known example: 'in the social production of their existence, men inevitably enter into definite relations, which are independent of their will, namely relations of production appropriate to a given stage in the development of their material forces of production' (Marx, 1972 [1859], p. 20).

As various commentators point out, however, on closer reading things are not so clear-cut. Two main arguments can be made that cast doubt on this simple interpretation. First, a finer reading of the works in which these quotes appear challenges their usual but more superficial interpretation.[1] Second, and more interestingly, it is clear that technological determinism sits very uncomfortably with the rest of Marx's contributions. For one thing, according to Marx, the emergence of capitalism seems to have had little to do with technology; the prime condition seems to have been the changing opportunities for profit making and in particular in response to the opening up of new markets.[2] Where this argument is given in great detail (e.g. in the Communist Manifesto) no mention is made of technological innovations at all. And on the occasions where Marx does talk of technical innovations (such as in navigation or in steam engines) these are understood as responses to the changed conditions of profit making.

An argument that Marx does make, which may be interpreted as encouraging the view of him as some form of technological determinist, is that sequence matters. Marx is conveying in the hand-mill and productive forces quotes the idea that some things cannot happen without other things having happened first. It is hard to see how the light bulb could have been designed before electricity, engine cooling

---

[1] For example, the hand-mill quotation can be read as a rather aphoristic 'heat of the moment' quip against Proudhon, which actually signifies little more than the idea that the hand-mill presupposes a different division of labour from the steam mill (Rosenberg, 1976). And MacKenzie amongst others notes that the comments in the Preface suggest technological determinism only if forces of production are equated with technology. But this is clearly not how Marx intends things, skills, experience, etc., of labour also being included (MacKenzie, 1984). For an interesting alternative view, see Shaw (1979).

[2] Following such events as the discovery of America, the opening up of East-Indian and Chinese markets, trade with colonies, etc.

systems before engines, etc.[3] By itself, an insistence on some idea of sequence is not particularly interesting, but this changes when combined with other elements of Marx's account that become clear if we focus on the second source of ambiguity or tension in Marx's work: whether technology is viewed as generally negative or positive.

Marx retained the Enlightenment idea that technology is ultimately progressive – a generator of change and probably necessary to create the material conditions for socialism to emerge (Kirkpatrick, 2008). Indeed, Marx's criticisms of capitalism are motivated not simply by the realisation that it is a miserable system to experience, but because the growth in technology, which should benefit all by creating vast new wealth, only ends up enriching a few and reduces the quality of life for most people. Thus technology should be making a positive impact on people's lives, but for most this is not the case. Marx's positive attitude to technology also shows through in his criticism of the Luddites' identification of machines as *the* problem and their attempts to destroy the machines they felt responsible for taking their jobs or undermining their skilled status. Moreover, although Marx was clearly critical of British imperialism, the justification of it, as bringing technological progress, was a position that Marx had some sympathy with – often bringing criticism of Marx as, amongst other things, Eurocentric (Adas, 1989, Baber, 1996).

However, Marx is perhaps better known for the less positive things he had to say about technology, especially in relation to his discussion of the way in which machines came to feature in the production process. Marx's account of the machine begins with the emergence of a class of propertyless wage labourers. For a variety of reasons, the norm of independent craftwork carried out by artisans was replaced by production carried out by employees, now working with tools (spinning wheels, looms, etc.,) belonging to a merchant, even though workers typically still worked in their own homes. Whilst the change in social relations was dramatic, however, the technical content of their work was largely unaltered.

Under such conditions the major route for increasing the surplus for merchants was through attempts to increase the working day. But this

---

[3] This is of course a trivial point, but it would seem that many so-called technological determinists have been labelled this way when some conception of sequence is all that is being argued for (another illustrative example is provided by the work of Ayres discussed in the previous chapter).

was not easily done. Thus, Marx argues, capitalists sought to control production more thoroughly, by bringing it into a physical space that could enable savings of fixed capital, ease of monitoring working effort, etc. Marx terms this stage of development 'manufacture', although the term is used literally to refer to the idea that production is still primarily by hand. Within the changes Marx considers, the crucial element is the division of labour. Marx focuses upon the general benefits of cooperation (or perhaps coordination or organisation would be a better word), but also upon their particular historical manifestation in a system in which workers, unable to perform or even understand the process of production as a whole, lose intellectual command over, and responsibility for, production.[4]

However, whilst manufacture generates increases in profits, Marx argues that such gains were severely limited because of the central role that human skills played in the process, a role that could be exploited by labour as the basis for different kinds of resistance. It is in this context that machines come into play. A machine, for Marx, is not simply a more complex tool, nor simply a systematic replacement of the labourer. Machines not only reduce the costs of the production process, but crucially discipline the existing workforce. Moreover, the alienation of the collective and intellectual aspects of work already diagnosed by Marx in simple cooperation and manufacture achieves a technical *embodiment* in the machine. In short, the emerging social relations of capitalism become materialised, concretised, or 'metalized' in the machine.[5]

These ideas of materialisation are central to those accounts that have taken seriously and attempted to extend Marx's work on technology. An illustrative example is the work of Harry Braverman (1975, see also Ihde, 1990). Whilst Braverman was centrally concerned with the way in which machine design reflected the needs of the owners of capital to control the labour process, he also highlighted what he terms the contradictory experience of technology. Whilst technical designs enabled the control over labour, the process whereby this control was realised largely went unnoticed and unchallenged. Engineers had internalised the value of cutting labour into the creation of the machines

---

[4] A good account of this is provided by MacKenzie (1984, p. 32).
[5] See for example Dyer-Witheford (1999, p. 40).

themselves, and for the workforce to oppose that technology appeared to go against common sense, even against 'progress'.

This last idea is clearly brought out by considering another development of Marx's position, namely, David Noble's account of machine tools production (Noble, 1984). This was one of the last areas of industrial production to be automated, leading to the redundancy of thousands of workers. When key decisions were made about introducing automation, managers and engineers had a choice between two automation methods. Both did similar jobs but in one, the automation was based upon copying the procedures undertaken by workers. Thus a role for the workforce still existed. In the other method, the automation process was based purely upon mathematical modelling techniques in which no such role for the workforce existed.

Noble's contribution is to show that although the two methods were functionally equivalent (indeed the former being marginally more efficient in the first instance), the latter was chosen because it would undermine the role of the workforce in future production. Once the technology was introduced, however, resisting it would appear as if resisting the very idea of 'progress'. Noble's work presents a clear picture of how on the one hand a particular technology (concerned with the automation of machine tools production) could embody values and effectively materialise the exploitative relationships that Marx had talked about, but appear beyond criticism. Thus, not only does technology serve to give permanence to relations of dominance but it also *legitimates* such relations. It is possible to say that ideas, values and relations determine the very structure of technology because, in the language of more recent accounts, technology is underdetermined by criteria of efficiency, functionality, etc.

To take stock, although Marx held that technology was a necessary condition for the development of socialism, he also identified the particular forms that technology took under capitalism as essential elements of a system that exploited and controlled the majority of the population. Fundamental here is Marx's emphasis upon the mechanisms whereby technology takes the form it does because of particular conflicts or activities, and in particular how technology features in the emergence and reproduction of particular social relations – the exploitative relations between owners of capital and labourers that Marx saw as central to understanding capitalism. Thus technology is the result of social activity that concretises or materialises particular values, class relations, etc., which then go on to become an important and enduring feature of the context within which further action takes place. In other words, whilst Marx maintains a position that is not deterministic, he does attribute a major role to technology in all manner of important social changes. He builds the idea of sequence and the appearance of inevitability of technology into an account of how particular technologies can be introduced legitimately, even though their actual implementation seems to be nothing other than attempts to control and exploit. In so doing, he provides a stark but plausible account of how particular technologies can be criticised, namely on the basis of their concretisation of particular relations of domination.

Before drawing out more implications from Marx's account, I wish to look briefly to an account of technology in which there is no ambiguity with respect to the thesis of technological neutrality, namely the work of Martin Heidegger.

Heidegger, Tool-Being and the Bad Life

It is generally accepted that Heidegger's contributions to the philosophy of technology are the most influential in the field (Scharff and Dusek, 2003). Heidegger is certainly the most prominent of a group of philosophers of technology who adopt a very dystopian view of technology.[6] Rather than attribute the problems of modernity to capitalism, as Marx did, these writers argued that the increasing prominence of technology is the primary problem of modern societies.

Heidegger's work is notoriously difficult to read, in good part because it involves as much obscure wordplay as logical argument. However, it is hard to find serious contributions to the history, sociology or philosophy of technology that do not refer (positively) to Heidegger's work. Arguably, one of the most consistent components in Heidegger's writings is his tool-analysis, or analysis of tool-being. Indeed, recent interpreters of Heidegger have even gone so far as to suggest that all Heidegger's main contributions are contained in his discussion of tool-being (e.g. Harman, 2009). The main message of this analysis is that the things that we interact with are never fully present in our consciousness, but rather are hidden or 'withdrawn realities' working in the background, largely unnoticed (Heidegger, 1962).

---

[6] See Chapter 1 for a discussion of other prominent examples.

To put this idea in context, Heidegger is breaking from his teacher Husserl, the founder of phenomenology, who focused on phenomena as they appear to consciousness, not 'things in themselves'. The actual workings of the (natural) world are bracketed for Husserl whilst we focus upon things as they appear to us, as they are experienced. Although it is fair to say that Husserl's position involves more than a simple form of empiricism (where the objects of the world are viewed as a bundle of qualities linked by the human mind), Husserl's project does suggest both a conception of the human agent as passively viewing the world and a world open to easy access. In contrast, and central to Heidegger's break with Husserl, Heidegger insisted that the correct model of the human agent is that of an active, working, *user* and the correct model of the world is one of complexity, much of which remains *hidden*. These differences are clearly illustrated by Husserl and Heidegger's criticisms of science. For Husserl, science tries to do too much, wrongly leaping beyond phenomena into a hidden world; for Heidegger, science is too shallow, wrongly reducing entities to measurable quantities.

Heidegger argues that given this 'withdrawal' of much of reality, it is a mistake to assume that we can perceive all there is. However, making such a mistake is not simply a matter of individual inadequacy or thoughtlessness. Rather, and in keeping with the central importance of 'practical doing' in Heidegger's work, different kinds of objects or aspects of reality reveal themselves in different ways through use, and thus encourage different ways of perceiving the world. Some aspects of reality (tool-being) provide hints or signals of the existence of different layers or complexities of reality. Here we have the famous examples of broken instruments, and natural objects, both of which Heidegger argues tends to provoke the perceiver to 'raise their head', so to speak, and wonder at their fuller reality. However, technology by contrast encourages a mistaken view of objects in isolation from everything else, hinting at nothing. These (technological) objects are simply taken for granted, things that fit seamlessly into networks of functional relations.

From this account of tool-being, Heidegger constructs a fully blown critique of modernity.[7] As technology becomes a more important component of modern societies, so the technological way of viewing

---

[7] For perhaps the best statement of this later development, see Heidegger (1977).

the world comes to dominate, thus giving rise to a series of social problems. In particular, not only are we encouraged to view the whole of our world in very instrumental terms, but through increased interaction with technology we are engaged in actually transforming the entire world (including ourselves) into stockpiles of resources or 'standing reserves', that is, objects to be controlled. Methodical and quantitative planning comes to dominate, destroying the 'integrity' of everything (Heidegger, 1977).

A good deal of the continued popularity of Heidegger's ideas comes from the fact that such images of modernity continue to resonate.[8] Of course similar images are invoked by a range of different writers, such as Weber's rationalisation or Habermas' invasion of the lifeworld (Weber, 1958, Habermas, 1984). But Heidegger's lasting relevance seems to lie in his insistence that it is our relationship with technology that holds the key to understanding the changes taking place. Moreover, technology's role is one that directly affects who or what we are, not simply the conditions we must negotiate. Crucial for Heidegger is the way that people come to see themselves. Diverse phenomena such as cosmetic surgery and the leisure industry are all examples, for Heidegger, of how we come increasingly to see ourselves as resources to be optimised.

Ultimately, for Heidegger, the result of increased technological advance is a kind of aimlessness, where the valued and important things in life are sacrificed in an automatic and passive pursuit of goals that are both unclear but somehow beyond challenge. For Heidegger, modernity is ultimately to be understood in terms of the replacement of meaning, creativity and the value of human life by a preoccupation with possession and control. As this replacement takes place, we become something else; we become isolated from the natural world, each other and even our own selves.

---

[8] We are surrounded by changes that seem to illustrate a growing instrumental mindset. Some of these were outlined in the literature on speeding up noted in the previous chapter. Common modern day examples include the damaging effects on healthcare of a preoccupation with cutting waiting lists, or on education by a preoccupation with league tables and exam results, or on the fragmentation of research communities due to the impossibility of being able to keep up with reading the massively increased output in academic papers produced by colleagues in response not to having more to say, but to pressures to meet quantitative criteria set by various research assessment exercises.

Against Technological Neutrality

I suggested above that the dismissal of technological determinism has led to a widespread, if rarely articulated, acceptance of some kind of technological neutrality. Whilst I do not wish to defend any form of technological determinism, I do want to argue against technological neutrality, or at least a particular form of it.

The position I am labelling as one of technological neutrality, sometimes also termed an instrumental conception of technology, has become commonplace in recent years, even if it is a position that is most often held only implicitly. The main idea is that there is nothing interesting or significant that can be said about technology itself. Technology is simply a neutral tool and any implications that follow from its use depend solely upon how that tool is used in a particular context.

As noted in Chapter 1, if such a position is taken seriously, then much of the philosophy of technology seems rather beside the point. The swings between positive and negative evaluations of technology, as well as the idea that technology is out of control and even the idea that technology might have particular roles to play in change, make little sense if technology is simply some kind of neutral tool. However, it does not follow that a rejection of a neutrality position requires an adherence to any form of determinism. Indeed, it does not follow that rejection of a neutrality position even involves a commitment to the idea that technology is generally either good or bad. Let me take these points in turn.

As noted at the beginning of this chapter, if humans have choice, any form of determinism must be rejected. Indeed, the social ontology outlined in Chapter 3 was in part formulated in order to explicitly avoid any kind of determinism. But it is a mistake to see a rejection of technological determinism as necessitating a commitment to the idea that technology has no properties or characteristics that can be described, analysed or explained. This is much like the (mistaken) idea that rejecting determinism involves giving up on any kind of causal explanation. In both cases, part of the problem is the implicit ontology at work, and in particular is the implicit acceptance of a constant conjunctions conception of causality.[9] If causality is a matter of constant conjunctions of events, it is difficult to see how determinism can be avoided without rejecting

---

[9] A prominent discussion of this is to be found in Bhaskar (1978); for more recent discussions see T. Lawson (1997).

any account of causal explanation. However, a structured ontology of mechanisms and powers that can be understood to be acting but unrealised or in existence but unactualised provides a far more nuanced conception of causality in which *tendencies* operate because of the ways things are organised and structured. Thus the realisation that particular causal mechanisms do not always have the same effect in practice does not make it impossible to formulate a causal account about them. The point is rather to identify general tendencies that operate because of a certain thing, such as technology, being the kind of thing that it is. Whether or not such tendencies actually result in particular events or states of affairs coming about is another issue and depends upon the particular mix of different causal mechanisms in play at any point in time.

Moving on to the second point, rejecting a neutrality position does not require a commitment to the idea that technology, in some general sense, is either good or bad. Whilst the philosophy of technology has involved many examples of such commitments, this is not my intention here. But neither is it the case that the position I am adopting is one in which only very specific, case-by-case, features of particular technologies can be discussed. Certainly there is much that can be said about particular instances of technology. Different technologies do different things: some connect us, some seem to separate us; some are good for the environment, some are not; some encourage democratic participation, some do not, etc. Langdon Winner's argument that nuclear technology (which requires vast resources spent on safety and security) is far less encouraging of democratic forms of organisation than say solar technology would be an example of this kind (Winner, 1980). However, the more interesting issue is whether or not we can say something more. Can we say something that is more general than such specific examples, but less general than 'all technology is good or bad'? It is this kind of 'middle level' causal tendency statement that I am concerned with here. And I want to suggest that both Marx and Heidegger give us useful resources in this respect. To make these arguments, however, I need to recast some of the arguments that Marx and Heidegger make in terms of the account of technology outlined above.

Taking Marx first, I have suggested that Marx's main contribution is to spell out the way in which technology comes to embody, or materialise, particular social relations. Two main features of this materialisation were focused upon above. First, it gives particular sets of power relations a form of permanency or endurability. Secondly, it also provides a source of legitimation of those power relations because the introduction of those machines, which embody a particular set of power relations, are viewed as the next step in the apparently inevitable progression of science, efficiency, etc.[10]

To these features, I wish to add one further idea. Periods in which such machines are introduced create moments of disruption to traditional ways of doing things, undermining some relations and practices and reinforcing others. As such we can interpret Marx as focusing in particular upon the positioning of technology. Although Marx does not formalise things in this way, he is also concerned with the idea, focused upon in Chapter 7, that technology extends human capabilities. In particular, technology is understood to extend the productive capabilities of the workforce, but also under capitalism the 'reach' of the owners of capital. It is this combination of ideas that lies at the heart of both increasing productivity or even of increasing the subservience of labour, and yet potentially under socialism not only increasing productive potentials but also a better standard of well-being for all.

The argument of Chapter 7 was that the idea of extension is most usefully understood in terms of networks of interdependencies or systems, which consist, at least in part, in the interconnection and interaction of people via the extension of their capabilities *through things*. As new devices are introduced, some people become more capable, some become less so, some become crucially important and

---

[10] Actually, these insights have been developed in two largely different directions, both of which were to some extent encouraged by Marx. Whilst these two points are consistent features of one direction, the other direction focuses upon the intrinsic problems with such a strategy for production, especially with the reasons for the declining rate of profit. Perhaps the best know example of this latter strategy is provided by Mandel (1975). Largely as a reaction to the post-industrialism of Bell and others, Mandell argued that increased technology could not offer a way of transcending the contradictions in capitalism grounded by Marx's wider analysis. This is essentially because, if labour is the source of surplus value, reducing the component of labour used in the production process (so increasing the organic composition of capital) ultimately must reduce the profitability of that production. Whilst it is easy to sympathise with Mandel's desire to challenge what can only be described as the triumphalism of much of the post-industrial literature of this time, it is not clear that Mandel incorporates anything more than a very technical and ultimately mechanistic understanding of how technology might operate in Marx's system (for a fuller discussion see Dyer-Witheford (1999)).

in a position of great power, some become less needed or redundant. Whatever the upshot, such moments are crucial, or ontologically significant, moments in which traditionally accepted, negotiated ways of doing things are disrupted and reconfigured.

Of course these different points can be combined. First, the nature of such potentially contested situations tend to be hidden by the fact that the introduction of technological devices is perceived as inevitable, the next step in the sequence, of progress, and thus beyond challenge. Second, criticism of the technologies being introduced, whilst still possible, is likely to require investigation of the kinds of relations that are built into and made legitimate by any particular technology.

However, it is possible to say more about these moments. In particular, Marx's account brings out the obvious implication that technologies require resources to be designed and developed, and the introduction of technologies will often be by those with the most resources and with particular positional interests. In short, it is not simply that technology can be criticised if it embodies and legitimises certain relations of domination, but it also lends itself to such outcomes. Such an argument of course can then be understood in contrast with those employed in the work of Ayres and the American institutionalists, discussed in Chapter 6, that technology continually provides an 'equalising' challenge to the existing vested interests, etc. But if the major source of disruption of existing ways of doing things are essentially technologies designed or selected by and for those with the resources to fund the process and turn the results to their benefit, then it might be expected that their intentions, values and positional interests will tend to dominate the form of the technology adopted and thus the possibilities for any such challenge.

None of this denies the fact that the introduction of new technology always offers the possibility of change at many different levels. This seems to be another factor that underlies Marx's rather longer-term perspective that technology offer's positive benefits for all. But, and in contrast to Ayres, Marx's account identifies a ratchet-like process in which moments of disruption and upheaval, following the introduction of some technology, can be used as moments in which to restructure or reorganise important segments of the social world, which will tend to happen, if unchallenged, in ways that benefit those with the greatest access to resources. It may be possible to combine the arguments of

Marx and Ayres here. Both highlight the tendency of new technologies to disrupt existing networks of interdependencies but focus upon different tendencies that are likely to be set in play following such disruptions. Either way, they both identify real causal tendencies arising from technology that also avoid any commitment to a simple (deterministic) account of how changes in technology will work out.

Let us now turn to Heidegger. Following from his tool-being analysis, Heidegger focuses upon the idea that 'being withdraws', and that this fact can be used to explain and characterise tendencies towards an increasingly isolated, functionalised and meaningless modernity. However, some of Heidegger's more persuasive arguments can be recast, perhaps less mysteriously, in terms of the account above. In particular, I want to suggest that it is possible to recast Heidegger's preoccupation with hiddenness and the technological 'enframing' in terms of ideas of open and closed systems and the distinction between methods of abstraction and isolation. It may seem that I am suggesting the potentially awkward replacement of an ontological position about the world with a distinction between epistemological approaches to knowing about the world. However, given Heidegger's phenomenological orientation, it is not clear that such clear distinctions are sustainable. Let me explain.

First, Heidegger's starting point, that nothing can be known about in its entirety, can be interpreted as being not so different from the idea that methods of abstraction are generally more appropriate to understanding the world than methods of isolation. To repeat, the crucial point about the method of abstraction is that it involves focusing on one particular aspect of reality but not forgetting that there is more to it; it is simply a matter of holding something in view, unlike the method of isolation that treats reality as though that aspect being focused upon is all there is, or at least that it is somehow insulated from the rest.[11] When we abstract, we are continually (intuitively) making decisions about how much of reality to focus upon and how much to leave out

---

[11] There is clearly much more than can be said about the relative merits or characteristics of abstraction and isolation than can be considered here. In particular, dealing with, or coming to understand, systems that are irreducibly open, as much of the social world seems to be, requires more than simple abstraction as I am presenting it here. For example, also important would be different forms of contrast explanation. For an extended account see for example see T. Lawson (1997, 2009, pp. 206–225).

(what should remain hidden), continually shifting the focus as the context requires. This of course, is exactly the kind of intuitive decision process that those with autism find so challenging. They prefer contextless 'whenever X then Y' rules to apply to the world (where nothing is hidden). But whereas such rules are appropriate when dealing with isolatable or differentiable features of the world (especially those aspects of technology that those with autism are most drawn to), much of reality is not differentiable, that is, it is not equally susceptible to understanding in terms of closed systems.

As long as there is some relatively isolatable mechanism that can be identified, and which can be understood to combine in relatively mechanistic ways, the method of isolation seems to be an appropriate method for coming to understand the world, and this, Heidegger is suggesting, seems to be encouraged by the increasing use of technology. However, where such differentiability does not exist, and so closures impossible to construct, these systems must be understood or made intelligible by some kind of (method of) abstraction. A difference here between Heidegger's idea of hiddenness seems to be that it is always a mistake to use methods of isolation; he does not distinguish between open and closed systems and so cannot admit that methods of isolation are sometimes appropriate. But there is a sense in which abstraction is always the more general approach and isolation is simply a special case. It would never be wrong to abstract (although it may be a slow way to come to understand the workings of some feature of reality). But it is a mistake to treat those aspects of reality that are intrinsically open or not differentiable (and for the social domain these must be the vast majority) as though they are closed or differentiable.

Such a mistake can be made at different levels. In Chapter 2, I argued that at the disciplinary level, most attempts at deductivist modelling in social sciences (as particularly evident in much of modern economics) also make this mistake – they rely upon or encourage the analysis of social systems in terms that are really only applicable in closed systems.[12] It was argued above that a tendency to deal with or view open systems as if they are closed also lies at the root of the problems those with autism face when engaging with, or understanding, social reality. Moreover, this inability was not thought to be restricted to a small number of children classically defined as autistic. Rather, given

---

[12] See T. Lawson (1997) for an extended discussion.

the spectrum nature of the condition (thus indicating natural differentiation rather than a simple impediment or well defined disease) it was argued that this tendency exists in different degrees throughout society, although it seems especially so amongst men.

Heidegger's attempt to generalise tool-being can thus be captured or recast in terms of a society-wide tendency towards the method of isolation rather than abstraction. And in this context, the distinctiveness of Heidegger's contribution comes in arguing that this particular way of viewing or understanding the world is encouraged by interacting with particular kinds of things, especially technology. As such, the discussion of autism above provides an interesting complement to Heidegger's vision of an impoverished modern society, which we might perhaps term, incorporating Heidegger's language, an 'autistic enframing'. Combining the understanding of autism, as laid out above, with Heidegger's analysis also provides a counterbalance to Heidegger's claim that it is the use of technology that causes all the problems. Rather, the use of technology is part of the problem, which can be understood to reinforce and select all kinds of personal predispositions, and abilities, as well.

Certain aspects of technology appeal to or are best understood in terms of closed systems. In particular, the coming into being of technology always involves a moment in which isolation is essential. To repeat, such ways of thinking are not in themselves problematic. Indeed they are most appropriate where isolatability really is possible. And in important aspects of technology (especially design and recombination) this is the case, as the relation of those with autism to technology suggests. In particular, it was noted that so many of those with autism are drawn to particular activities, especially those concerned with engineering, computer science, etc., which are primarily the activities focused upon in this isolative moment in technical activity.[13] What makes such phenomena understandable, I am suggesting, is the

[13] And it has even been argued that the main reason why so many economists persevere with methods of isolation in the social world is that they are themselves on the autism spectrum. The point being that activities that require and encourage methods of isolation, the use of mathematical modelling being an obvious example, are attractive to those who are most comfortable with methods of isolation, even if it seems to be the case that such methods are of limited use. See for example the website 'A Brief History of the Post-Autistic Economics Movement' at www.paecon.net/HistoryPAE.htm, and J. Lawson (2013).

importance of the separation of particular causal mechanisms which are then recombined to make particular technologies.

But even if some aspects of technology might encourage such a way of viewing or relating to the world, why should this diffuse to become a society-wide general orientation, capable of explaining many of the ills of modernity, as Heidegger suggests? In fairness, Heidegger remained quite vague about the mechanisms here, usually referring to the superficial treatment of reality by science. But combining the accounts above, it would seem that a variety of different mechanisms seem to be open to use. In particular, if we grant, for a moment, that there are real tendencies in the direction that Heidegger suggests, which would of course explain the continued interest in Heideggerian themes as noted above, then some of the following may be brought to bear.

One possible route seems to be via design. As was argued at the end of Chapter 9, those who are drawn to and show themselves to be competent at isolating (methods of isolation) are drawn to design, and indeed many of the technological devices we encounter are designed by those with this orientation. These devices then appear, and are adopted, as efficient, as the latest fashion, etc. In other words, our technologies are increasingly designed by those who may well find difficulty in dealing with distinctly social aspects of reality. This was the argument advanced in relation to calls for further democratisation of technology. But viewed in the context of Heidegger's contributions, design by those for whom abstraction is a problem introduces a bias towards the production of objects and devices that encourage a mechanical or instrumental world view. Given the way that new technologies appear as simply the currently most efficient solution or device, they effectively act as a kind of Trojan horse bringing with them isolative values or conceptions of the world as differentiable along with their adoption.

Another mechanism may be that by which those drawn to the isolative moment become icons or heroes of a generation in which technology comes to dominate. For example, note the spate of films in the early twenty-first century about those such as Steve Jobs, Mark Zuckerberg and Bill Gates who, through the successful design and marketing of new technology, have become successful, all world-changing individuals, very wealthy and famous, but also appear to be firmly on the spectrum. In previous generations, each may have been viewed as 'odd', on the margins of society, but in a society in which technology is so central, they are successes – not only tolerated but emulated.

Yet another mechanism may be the way in which we attempt to explain the world. By this I am primarily drawing attention to the importance of retroduction or abduction in explanation, as discussed in Chapter 4. The important point here is that much of the time we explain things, in contrast to what philosophers of science might have us believe, neither by induction nor deduction. Rather we use metaphor and analogy, bringing mechanisms known about in one domain to another domain to generate explanations of the kind, if X were to exist, might Y follow? Explanation, understood in this way (as retroduction), is particularly dependent upon the kinds of analogies and metaphors at the disposal of a particular community. If that community is one that spends most of its time interacting with technology such as computers, machine tools or whatever, there is at least scope for the pervasiveness of mechanistic metaphors based upon the not inappropriate methods of isolation that emerge in interacting with technology, to be increasingly drawn upon when dealing with other, less appropriate aspects of reality.

There is little space here to be anything other than suggestive, but it does seem that many of the ideas for which Heidegger is best known can be usefully accommodated in these terms. This seems particularly so for those aspects of Heidegger's work most taken up and extended by others. For example, one especially prominent extension of Heidegger's analysis is provided by Albert Borgmann (1984). Although Borgmann is critical of Heidegger for what he sees as pessimistically 'harking back to the simpler life of yesteryear', it is clear that Borgmann is concerned with many of the same issues as Heidegger and indeed arrives at many of Heidegger's major conclusions.

The main appeal of technology, Borgmann suggests, is that it offers gains in efficiency, but that these come at the cost of distancing users from reality. This distance emerges because technology separates off the thing (quality, good) it delivers from the context and means of its delivery (heat appears without burning anything, email arrives without personal contact, etc.,). This leads Borgmann to distinguish between a device (a mass-produced technological artefact) and a focal thing (an artefact that reflects natural or social context). Whereas a focal thing requires patience, skill, attention, effort, etc., and locates us in a place and gives us a sense of identity, a device does none of these things.

Consider, for example, the replacement of a wood-burning stove by central heating. Both provide warmth, and so are in a sense functionally equivalent. But the former requires that the family be far more in touch with the processes at work. They are called upon to look for wood, perhaps build a shed to store the wood, congregate together around it to enjoy the warmth, etc. Central heating, in contrast, requires or facilitates none of these things. Whereas the wood-burning stove brings people together, prompts them to invest in skills and a sense of their own place, central heating indeed tends to undermine all of these. The influence of Heidegger is very clear, Borgmann providing a novel recasting of Heidegger's distinction between the technological 'enframing' and the power of the traditional or cultural to 'gather' people and nature together.

Another problem, according to Borgmann, is that although technology is adopted because it promises to unburden us from unnecessary daily tasks, so allowing us to do more important things, in reality all it tends to free us for is the passive consumption of more technology. For example, greater processing speed on our computers does not free us to go and paint pictures or play music – we rather spend more time on the computer. For Borgmann, the more that we use technology the more we become disengaged consumers only interested in more devices. Thus we end up with another familiar Heideggerian conclusion: activities are drained of meaning and the values of possession and control tend to dominate.

Borgmann's analysis is most often met by two interrelated criticisms. The first is that the emphasis Borgmann places on losing sight of how things happen, on how technologies tend to 'put things under the bonnet', is overstated. Instead, it is argued, such ideas are not confined to technology as Borgmann suggests, and much in the social world takes this form, such as in our development of tacit skills, our reliance on social institutions, etc. Here, our knowledge about the world becomes embodied in all kinds of things (not just technology), but also into habits, routines, conventions, markets, etc. In large part, this embodiment takes place so that we can get on with dealing with more immediate concerns; it is impossible to hold all the information necessary for our interactions with the world in our discursive consciousness, and there are many ways of putting this information 'under the bonnet' in Borgmann's sense. However, I would argue that the significance of this process where technology is concerned is the degree to which once this knowledge becomes embodied in technological artefacts, it does not constantly need to recycle through a discursive moment in the minds of the users. Indeed increasingly it cannot. Much that is tacit for us is returned to whenever we come across particular conflicts or surprises. But for technology, this 'under the bonnet-ness' refers primarily to the way that things can be atomistically combined in such a way that they do not require more thought. Given the increasingly complex and integrated nature of technological objects, as brought out in Simondon's work above, we simply do not cycle through discursive moments as we do with routines and rules, etc. Rather when we face problems with our technologies, we send them to experts to fix, or increasingly, we simply discard them in favour of newer models. To investigate under the bonnet is an activity that increasingly is reserved for technicians and other experts. The user then tends to feel increasingly more dependent upon technological experts and more uncertain about and indeed more alienated from just the technological artefacts they wish to use.

Another criticism of Borgmann's work has been the problem of providing a clear demarcation between those artefacts that are considered to be devices and those artefacts considered to be focal things (Higgs et al., 2000). In particular, is it not the case that some devices have the potential to become focal? That the Amish use some of the most cutting-edge, high-technology barbeques in the world would seem to suggest just this (Kraybill et al., 2013). And indeed, in different situations can't the same technological artefact have the properties of both a device and a focal thing? If Borgmann agrees that this is possible, of course, it appears that the problem may not then be technological at all, it is rather what we do with our technologies that matters. Thus we arrive back at a kind of neutrality view that Borgmann is keen to avoid.

However, the above account appears to be able to accommodate much of Borgmann's argument without ending up in a neutrality position. But let us approach the ideas slightly differently. Rather than start from a position in which there are different kinds of artefacts (focal things and devices – 'good' and 'bad' artefacts) let us start from the idea, suggested above that all artefacts, to be made to work, must be enrolled or positioned. For Borgmann, an important aspect of this enrolment process is that it is a means by which knowledge of the 'good-life' comes to be embedded into artefacts or, more generally, into traditional ways of living which are given permanence or endurability by the enrolment or positioning of different objects and artefacts into our lifeworlds. We adopt new technologies, in Borgmann's sense, because we can see what some new device, in principle, might be able to do, and because we see a desire for such a 'saving' in our everyday lives. But adopting a new technology and enrolling it, both embeds knowledge of the world and also positions us, the user, in a particular system of interdependencies through such use. For example, using a mobile in a particular place (train, library, etc.,) immediately inserts us into an ongoing, contested, set of relations revolving around the appropriate rules to follow in using mobiles. And existing intuitions or understandings of what makes for a better world will be part of the considerations taken into account by all concerned in the efforts to renegotiate these relations, or enrol devices such as mobile phones, enrolment taking forms that preserve those aspects of our lifeworld that we do not want to let go of. It might involve leaving phones out of our houses, as the Amish have insisted, or having phone-free pubs or carriages on trains. The point is that renegotiating the ways in which such devices are enrolled takes time, thought and effort if such enrolment is to embed our values or ideas of the good-life in the way that Borgmann suggests. In which case, the point to emphasise in Borgmann's analysis, and as was argued in the previous chapter, is that there may simply not be enough time.

In particular, new artefacts may arrive on the scene at a rate that is too fast for any 'healthy' form of enrolment to take place. And, for the reasons provided in the previous chapter, this is more likely to be the case for technological artefacts. Swayed by the promise of an easier life, or excited by the extended set of possibilities on offer, we buy into new technologies at a rate that cannot really be matched by our attempts to embedded such developments within the systems of relations in which we exist, embodying the lessons that past generations have learned about the good-life. In this case, we may not have to insist on such a severe distinction between devices and things, as Borgmann seems to suggest. Rather, all artefacts, even technological artefacts, have the potential to become focal things. But it can be argued that it is a feature of technologies that they tend to appear at ever faster rates. There will come a point where the arrival of new technologies outpaces the ability of users to enrolled them into their lifeworld in healthy and meaningful ways. Formulated in this way, Borgmann's contribution becomes an essentially communitarian argument based both on the modes of being of technological artefacts and of social relations involved in their positioning or enrolment (the latter being slower to form and more precarious to reproduce). As such, we have effectively arrived back at arguments made specifically about the tendencies towards speeding up and trivialisation of modern societies discussed in the previous chapter. But now, by drawing upon the contributions of those such as Heidegger and Borgmann, there is far more complexity to this argument and obvious scope for further development.

Complementarities

The main aim of this chapter has been to challenge the idea of techno-logical neutrality and suggest the kinds of features that we might expect from our day-to-day interactions with technology. In short, the aim has been to suggest that a rejection of technological determinism does not necessitate a commitment to the idea that technology has no general properties or characteristics that can be described, analysed and explained. The discussion of Marx and Heidegger, prominent exam-ples of contributors identified as holding some kind of technological determinism, has been undertaken to show how their concerns, rather than constituting any form of determinism, are with general tendencies that might be associated with technology because of the kind of thing it is. Marx and Heidegger, I have argued, provide us with ideas, insights, concerns and observations about the general properties of technology that can be used as resources to generate an account of real tendencies that do operate, irrespective of whether they are manifest in any parti-cular events or outcomes. In order to draw out such features, I have been necessarily selective in my focus upon the work of both authors, and taken various liberties in the interpretation I have offered. However, the point is to show that they are accounts that avoid both determinism and neutrality.

I do also believe that once recast in the terms above, certain complementarities between the ideas of Marx and Heidegger come to light, at least with regard to understanding the nature of technology. Marx's account, especially understood in terms of the extension of human capabilities and significant ontological moments, can explain much of the disruption that new technology brings to existing sets of relationships, conventions, rules, etc. If we add to this the arguments about recombination and its role in the exponential growth in devices, it is easy to see why new technologies come along at a pace which are increasingly difficult to enrol in meaningful or healthy ways, thus contributing to a general perception of the world in terms of differentiable or closed systems, something, which borrowing from Heidegger we might term some kind of 'autistic enframing'.

Moreover, we can view the work of Heidegger and Marx as concerned with different moments in the coming into being of technology, Heidegger focusing upon the moment of isolation and Marx, for the most part, focusing upon the enrolment or positioning of technology. In terms of the ongoing use of technology, viewed in the kind of recursive or transformational manner suggested in Chapter 3, both Marx and Heidegger contribute to an understanding of technology as condition and consequence of technical activity. Technical activity is conditioned not only by the technological enframing suggested by Heidegger, but by the relations of domination materialised in particular technologies. Similarly, the consequences of technical action not only lead to particular materialisations of the relations under which such action is engaged, but in producing particular technologies that are only superficially embedded, a more trivial relation to the world is likely to be reproduced.

From the vantage point of the conception of technology set out above, it is then possible to suggest that the two authors' writings, on technology at least, are not only complementary but are each in need of (something like) the other's position. Moreover, it seems possible to say that both Marx and Heidegger identify mutually reinforcing tendencies and processes that fare particularly well under capitalism, each providing the conditions for success of the other or contributing directly one to the other.

The more general point, I want to suggest, is that once the emphasis is placed upon the mix of more and less isolatable features of any particular technology, it is possible to rethink many of the traditional concerns of the philosophy of technology, including those of Marx and Heidegger, in new and potential fruitful ways. This focus upon the work of Marx and Heidegger has been adopted both because of the dominant and contentious positions both occupy within the literature, and also because I believe they still have much to say. But I believe there are many other contributors to the philosophy of technology who's work can be revisited usefully in this way. Indeed, the main motivation of this book has been to try to put some of these older debates back onto the agenda. Such a project is required not only because these debates hold much of use, but because they have been rejected and dismissed for reasons that seem misguided. Thus there has been the need not only to revisit these older ideas but to reformulate and recast them in the light of new ideas, especially those emerging from a concern with ontology. At the very least, the aim has been to encourage dialogue between those interested in technology and those interested in ontology. There is certainly no denying that much of the work has yet to be done, but the focus of this book upon the relationship between technology and isolation seems to be fundamental, and a good place to start.

Bibliography

Abramovitz, M. (1956). Resource and output trends in the United States since 1870. *American Economic Review*, vol. 46, no. 2, 5–23.

(1989). *Thinking about growth and other essays on economic growth and welfare*. Cambridge & New York, Cambridge University Press.

Adam, B. (1990). *Time and social theory*. Oxford, Polity.

Adas, M. (1989). *Machines as the measure of men: science, technology, and ideologies of Western dominance*. Ithaca, NY, Cornell University Press.

Adrien, J.L., E. Ornitz, C. Barthelemy, D. Sauvage & G. Lelord. (1987). The presence or absence of certain behaviors associated with infantile autism in severely retarded autistic and nonautistic retarded children and very young normal children. *Journal of Autism and Developmental Disorders*, vol. 17, no. 3, 407–416.

Akrich, M. (1992). The de-scription of technical objects. In *Shaping technology/building society*, ed. W.E.L. Bijker, J., 205–224. London, MIT Press.

Allison, A., J. Currall, M. Moss & S. Stuart. (2005). Digital identity matters. *Journal of the American Society for Information Science and Technology*, vol. 56, no. 4, 364–372.

Anscombe, G.E.M. (1971). *Causality and determination: an inaugural lecture*. London, Cambridge University Press.

Antonelli, C. (2008). *The economics of innovation: critical concepts in economics*. London & New York, Routledge.

Appadurai, A. (2001). *Globalization*. Durham, NC & London, Duke University Press.

Archer, M.S. (1995). *Realist social theory: the morphogenetic approach*. Cambridge & New York, Cambridge University Press.

Arrow, K.J. (1962). The economic implications of learning by doing. *Review of Economic Studies*, vol. 29, no. 80, 155–173.

(1969). Classificatory notes on production and transmission of technological knowledge. *American Economic Review*, vol. 59, no. 2, 29–35.

Arthur, W.B. (2009). *The nature of technology: what it is and how it evolves.* New York, Free Press.

Åsberg, C. & N. Lykke. (2010). Feminist technoscience studies. *European Journal of Women's Studies*, vol. 17, no. 4, 299–305.

Ayres, C.E. (1978 [1944]). *The theory of economic progress.* Kalamazoo, MI, New Issues Press.

Baber, Z. (1996). *The science of empire: scientific knowledge, civilization, and colonial rule in India.* Albany, State University of New York Press.

Bacon, F. (1909). *New Atlantis.* Cambridge, The University Press.

Baltaxe, C.A.M. (1977). Pragmatic deficits in the language of autistic adolescents. *Journal of Pediatric Psychology*, vol. 2, no. 4, 176–180.

Barad, K. (2003). Posthumanist performativity: toward an understanding of how matter comes to matter. *Signs*, vol. 28, no. 3 (2003): 801–831.

Barnbaum, D.R. (2008). *The ethics of autism: among them, but not of them.* Bloomington, Indiana University Press.

Barnes, B. (1983). Social life as bootstrapped induction. *Sociology*, vol. 17, no. 4, 524–545.

(1988). *The nature of power.* Cambridge, Polity Press.

Baron, R.M. & L.A. Boudreau. (1987). An ecological perspective on integrating personality and social psychology. *Journal of Personality and Social Psychology*, vol. 53, 1222–1228.

Baron-Cohen, S. (2004). *The essential difference: male and female brains and the truth about autism.* New York, Basic Books.

(2009). Autism: the empathizing-systemizing (E-S) theory. *Annals of the New York Academy of Science*, vol. 1156, 68–80.

(2016). Autism, maths, and sex: the special triangle. *The Lancet Psychiatry*, vol. 2, no. 9, 790–791.

Baron-Cohen, S., B. Auyeung, B. Nørgaard-Pedersen, D.M. Hougaard, M.W. Abdallah, L. Melgaard, A.S. Cohen, B. Chakrabarti, L. Ruta & M.V. Lombardo. (2015). Elevated fetal steroidogenic activity in autism. *Molecular Psychiatry*, vol. 20, no. 3, 369–376.

Baron-Cohen, S., A.M. Leslie & U. Frith. (1985). Does the autistic child have a theory of mind? *Cognition*, vol. 21, no. 1, 37–46.

Baron-Cohen, S., M.V. Lombardo, B. Auyeung, E. Ashwin, B. Chakrabarti & R. Knickmeyer. (2011). Why are autism spectrum conditions more prevalent in males? *PLOS Biology*, vol. 9, no. 6, e1001081.

Baron-Cohen, S., S. Wheelwright, J. Lawson, R. Griffin & J. Hill. (2002). The exact mind: empathising and systemising in autism spectrum conditions. In *Handbook of cognitive development*, ed. U. Goswami, 491–508, Oxford, Blackwell.

Bauman, M.L. & T.L. Kemper. (2005). *The neurobiology of autism.* Baltimore, Johns Hopkins University Press.

Bellini, S. & J. Akullian. (2007). A meta-analysis of video modeling and video self-modeling interventions of children and adolescents with autism spectrum disorders. *Exceptional Children*, vol. 73, no. 3, 264–287.

Bennett, J. (2010). *Vibrant matter: a political ecology of things*. Durham, NC, Duke University Press.

Bernard-Opitz, V., A. Chen, A.J. Kok & N. Sriram. (2000). Analysis of pragmatic aspects of communication behavior of verbal and nonverbal autistic children. *Prax Kinderpsychol Kinderpsychiatr*, vol. 49, no. 2, 97–108.

Bernard-Opitz, V., N. Sriram & S. Nakhoda-Sapuan. (2001). Enhancing social problem solving in children with autism and normal children through computer-assisted instruction. *Journal of Autism and Development Disorders*, vol. 31, no. 4, 377–384.

Berry, D.M. (2011). *The philosophy of software: code and mediation in the digital age*. Basingstoke, Hampshire & New York, Palgrave Macmillan.

Bhaskar, R. (1978). *A realist theory of science*. Brighton, Harvester.

(1986). *Scientific realism and human emancipation*. London, Verso.

(1989). *The possibility of naturalism*. Brighton, Harvester.

Bijker, W.E. (1987). The social construction of Bakelite: towards a theory of invention. In *The social construction of technological systems: new directions in the sociology and history of technology*, eds. W. Bijker, T. Pinch & T. Hughes. Cambridge, MA, MIT Press.

(1995). *Of bicycles, Bakelites, and bulbs: toward a theory of sociotechnical change*. Cambridge, MA, MIT Press.

Bijker, W., T. Pinch & T. Hughes. (1987). *The social construction of technological systems: new directions in the sociology and history of technology*. Cambridge, MA, MIT Press.

Bimber, B. (1996). The three faces of technological determinism. In *Does technology drive history?*, eds. M. Smith & L. Marx, 79–100. Cambridge, MA, MIT Press.

Binford, L.R. (1962). Archaeology as anthropology. *American Antiquity*, vol. 28, no. 2, 217–225.

Bloor, D. (1976). *Knowledge and social imagery*. London, Routledge.

(1997). *Wittgenstein, rules and institutions*. London, Psychology Press.

(2001). Wittgenstein and the priority of practice. In *The practice turn in contemporary theory*, eds. T. Schatzki, K. Knorr-Cetina & E. von Savigny, 95–106. London, Routledge.

Böhme, G. (2012). *Invasive technification: critical essays in the philosophy of technology*. London, Bloomsbury Publishing.

Boon, M. (2004). Technological instruments in scientific experimentation. *International studies in the philosophy of science*, vol. 18, no. 2–3, 221–230.

(2015). The scientific use of technological instruments. In *The role of technology in science: philosophical perspectives*, ed. S.O. Hansson, 55–79. Dordrecht, Springer.

Borgmann, A. (1984). *Technology and the character of contemporary life.* Chicago, IL, University of Chicago Press.

Bosch, G. (1970). *Infantile autism.* New York, Springer-Verlag.

Bourdieu, P. (1977). *Outline of a theory of practice.* Cambridge, Cambridge University Press.

(1990). *The logic of practice.* Stanford, CA, Stanford University Press.

Brassier, R., I.H. Grant, G. Harman & Q. Meillassoux. (2007). Speculative realism. *Collapse*, vol. 3, 306–449.

Braverman, H. (1975). *Labor and monopoly capital: the degradation of work in the twentieth century.* New York, Monthly Review Press.

Brey, P. (1997). Social constructivism for philosophers of technology: a shopper's guide. *Techné: Journal of the Society for Philosophy and Technology*, vol. 2, 56–78.

(2000). Theories of technology as extension of the human body. In *Research in philosophy and technology*, vol. 19, ed. C. Mitcham, 59–78. New York, JAI Press.

Brinkman, R. (1997). Toward a culture-conception of technology. *Journal of economic issues*, vol. 31, no. 4, 1027–1038.

Bronet, F. & L.L. Layne. (2010). Teaching feminist technology design. In *Feminist technology: women, gender, and technology*, eds. L.L. Layne, S.L. Vostral & K. Boyer, 197–202. Urbana, University of Illinois Press.

Bryant, L.R. (2011). *The democracy of objects.* Ann Arbor, Open Humanities Press.

(2012). Posthuman technologies. *Umbr(A)*, vol. 1, 25–41.

(2014). *Onto-cartography: an ontology of machines and media.* Edinburgh, Edinburgh University Press.

Bunge, M. (1966). Technology as applied science. *Technology and Culture*, vol. 7, no. 3, 329–347.

Burkhardt, H. & B. Smith. (1991). *Handbook of metaphysics and ontology.* Munich & Philadelphia, Philosophia Verlag.

Callon, M. (1987). Society in the making: the study of technology as a tool for sociological analysis. In *New directions in the sociology and history of technology*, eds. W. Bijker, T. Pinch & T. Hughes. Cambridge, MA, MIT Press.

Cartwright, N. (1999). *The dappled world: a study of the boundaries of science.* Cambridge & New York, Cambridge University Press.

Chapin, P.S. (1928). *Cultural change.* New York, The Century Co.

Cockburn, C. & S. Ormrod. (1993). *Gender and technology in the making.* London & Thousand Oaks, CA, Sage.

Colby, K.M. (1973). Rationale for computer-based treatment of language difficulties in nonspeaking autistic children. *Journal of Autism and Childhood Schizophrenia*, vol. 3, no. 3, 254–260.

Collier, A. (2003). *In defence of objectivity and other essays: on realism, existentialism and politics*. London & New York, Routledge.

Collins, H.M. (1985). *Changing order: replication and induction in scientific practice*. Beverly Hills, California, Sage.

(2001). What is tacit knowledge? In *The practice turn in contemporary theory*, eds. T Schatzki, K. Knorr-Cetina & E. von Savigny, 107–119. London, Routledge.

Connolly, W.E. (2013a). *The fragility of things: self-organizing processes, neoliberal fantasies, and democratic activism*, Durham, Duke University Press.

(2013b). The 'new materialism' and the fragility of things. *Millennium-Journal of International Studies*, vol. 41, no. 3, 399–412.

Constant, E.W. (1984). Communities and hierarchies: structure in the practice of science and technology. In *The nature of technological knowledge. Are models of scientific change relevant?*, 27–46. Netherlands, Springer.

Coole, D. (2013). Agentic capacities and capacious historical materialism: thinking with new materialisms in the political sciences. *Millennium: Journal of International Studies*, vol. 41, no. 3, 451–469.

Coole, D.H. & S. Frost. (2010). *New materialisms: ontology, agency, and politics*. Durham NC & London, Duke University Press.

Cooper, R.A., K.C. Plaisted-Grant, S. Baron-Cohen & J.S. Simons. (2016). Reality monitoring and metamemory in adults with autism spectrum conditions. *Journal of Autism and Developmental Disorders*, vol. 46, no. 6, 2186–2198.

Corneliussen, H.G. (2014). Making the invisible become visible: recognizing women's relationship with technology. *International Journal of Gender, Science and Technology*, vol. 6, no. 2, 209–222.

Courchesne, E., P.R. Mouton, M.E. Calhoun, K. Semendeferi, C. Ahrens-Barbeau, M.J. Hallet, C.C. Barnes & K. Pierce. (2011). Neuron number and size in prefrontal cortex of children with autism. *JAMA*, vol. 306, no. 18, 2001–2010.

Cowan, R.S. (1976). The Industrial Revolution in the home: household technology and social change in 20th century. *Technology and Culture*, vol. 17, no. 1, 1–23.

Dautenhahn, K. (2003). Roles and functions of robots in human society: implications from research in autism therapy. *Robotica*, vol. 21, 443–452.

DeMyer, M.K., J.N. Hingtgen & R.K. Jackson. (1981). Infantile autism reviewed: a decade of research. *Schizophrenia Bulletin*, vol. 7, no. 3, 388–451.

DeSanctis, G. & M.S. Poole. (1994). Capturing the complexity in advanced technology use: adaptive structuration theory. *Organization science*, vol. 5, no. 2, 121–147.

Dosi, G. & M. Grazzi. (2010). On the nature of technologies: knowledge, procedures, artifacts and production inputs. *Cambridge Journal of Economics*, vol. 34, no. 1, 173–184.

Dusek, V. (2006). *Philosophy of technology: an introduction*. Malden, MA & Oxford, Blackwell Pub.

Dyer-Witheford, N. (1999). *Cyber-Marx: cycles and circuits of struggle in high technology capitalism*. University of Illinois Press.

Ekbia, H.R. (2009). Digital artifacts as quasi-objects: Qualification, mediation, and materiality. *Journal of the American Society for Information Science and Technology*, vol. 60, no. 12, 2554–2566.

Elder-Vass, D. (2007). For emergence: refining Archer's account of social structure. *Journal for the Theory of Social Behaviour*, vol. 37, no. 1, 25–44.

(2010). The emergence of culture. G. Albert & S. Sigmund, eds. *Sociological Theory Controversial, 50th Special Edition of the Cologne Journal of Sociology and Social Psychology*, 351–363.

Ellis, B.D. (2001). *Scientific essentialism*. Cambridge & New York, Cambridge University Press.

Ellul, J. (1964). *The technological society*. New York, Knopf.

(1980). *The technological system*. New York, Continuum.

Ennis, P. (2011). *Continental realism*. Zero Books.

Fatemi, S.H. (2016). *The molecular basis of autism*. London, Springer.

Faulkner, P., C. Lawson & J. Runde. (2010). Theorising technology. *Cambridge Journal of Economics*, vol. 34, no. 1.

Faulkner, P. & J. Runde. (2009). On the identity of technological objects and user innovations in function. *Academy of Management Review*, vol. 34, no. 3, 442–462.

(2013). Technological objects, social positions, and the transformational model of social activity. *MIS Quarterly*, vol. 37, no. 3, 803–818.

Feenberg, A. (2000). *Questioning technology*. New York, Routledge.

(2001). Democratizing technology: interests, codes, rights. *The Journal of Ethics*, vol. 5, no. 2, 177–195.

(2002). *Transforming technology: a critical theory revisited*. New York, Oxford University Press.

(2010). *Between reason and experience: essays in technology and modernity*. Cambridge, MA, MIT Press.

(2014). *The philosophy of praxis: Marx, Lukács, and the Frankfurt school*. London, Verso.

Feibleman, J.K. (1979). Technology and human nature. *Southwestern Journal of Philosophy*, vol. 10, no. 1, 35–41.

Fine, A. (1986). *The shaky game: Einstein, realism, and the quantum theory*. Chicago, IL, University of Chicago Press.

Floris, D.L., M.-C. Lai, T. Auer, M.V. Lombardo, C. Ecker, B. Chakrabarti, S.J. Wheelwright, E.T. Bullmore, D.G.M. Murphy, S. Baron-Cohen & J. Suckling. (2016). Atypically rightward cerebral asymmetry in male adults with autism stratifies individuals with and without language delay. *Human Brain Mapping*, vol. 37, no. 1, 230–253.

Forman, P. (2007). The primacy of science in modernity, of technology in postmodernity, and of ideology in the history of technology. *History and Technology*, vol. 23, no. 1–2, 1–152.

Franssen, M.E., P.E. Kroes, T.A.C.E. Reydon & P.E.E. Vermaas. (2014). *Artefact kinds: ontology and the human-made world*. Switzerland, Springer.

Freeman, B.J., E.R. Ritvo & P.C. Schroth. (1984). Behavior assessment of the syndrome of autism – behavior observation system. *Journal of the American Academy of Child and Adolescent Psychiatry*, vol. 23, no. 5, 588–594.

Frith, U. (1989). A new look at language and communication in autism. *British Journal of Disorders of Communication*, vol. 24, no. 2, 123–150.

(2003). *Autism: explaining the enigma*. Malden, MA; Oxford, Blackwell Publishing.

Frith, U. & F. Happé. (1994). *Autism: beyond 'theory of mind'*. Cognition, vol. 50, no. 1–3, 115–132.

Fullbrook, E. (2009). *Ontology and economics: Tony Lawson and his critics*. New York, Routledge.

Gehlen, A. (1980). *Man in the age of technology*. New York, Columbia University Press.

Geschwind, D.H. (2008). Autism: many genes, common pathways? *Cell*, vol. 135, no. 3, 391–395.

Geschwind, D.H. & P. Levitt. (2007). Autism spectrum disorders: developmental disconnection syndromes. *Current opinion in neurobiology*, vol. 17, no. 1, 103–111.

Gibson, J.J. (1977). The theory of affordances. In *Perceiving, acting, and knowing: toward an ecological psychology*, eds. R. Shaw & J. Bransford. Hillsdale, NJ, Lawrence Erlbaum.

(1986). *The ecological approach to visual perception*. Hillsdale, NJ, Lawrence Erlbaum.

Gibson, E.J. & A.D. Pick. (1979). *Perception and its development: a tribute to Eleanor J. Gibson.* New York, L. Erlbaum Associates.

Giddens, A. (1984). *The constitution of society: outline of the theory of structuration.* Cambridge, UK, Polity Press.

Giddens, A. & C. Pierson. (1998). *Conversations with Anthony Giddens: making sense of modernity.* Stanford University, Stanford University Press.

Gilfillan, S.C. (1935a). *Inventing the ship.* Chicago, IL, Follett Publishing Company.

(1935b). *The sociology of invention; an essay in the social causes of technic invention and some of its social results.* Chicago, IL, Follett Publishing Company.

Gillberg, C., S. Ehlers, H. Schaumann, G. Jakobsson, S.O. Dahlgren, R. Lindblom, A. Bagenholm, T. Tjuus & E. Blidner. (1990). Autism under age 3 years – a clinical-study of 28 cases referred for autistic symptoms in infancy. *Journal of Child Psychology and Psychiatry and Allied Disciplines*, vol. 31, no. 6, 921–934.

Gillespie, C.C. (1960). The edge of objectivity. *An Essay in the History of Scientific Ideas. Princeton*, vol. 52, no. 3, Princeton, NJ, Princeton University Press.

Gispen, K. (1989). *New profession, old order: engineers and German society, 1815–1914.* Cambridge, Cambridge University Press.

Gleick, J. (1999). *Faster: the acceleration of just about everything.* New York, Pantheon Books.

Gluer, K. & P. Pagin. (2003). Meaning theory and autistic speakers. *Mind & Language*, vol. 18, no. 1, 23–51.

Godin, B. (2010). Innovation without the word: William F. Ogburn's contribution to the study of technological innovation. *Minerva*, vol. 48, no. 3, 277–307.

Goldsmith, T.R. & L.A. LeBlanc. (2004). Use of technology in interventions for children with autism. *Journal of Early and Intensive Behavior Intervention*, vol. 1, no. 2, 166–178.

Graham-Rowe, D. (2002). My best friend's a robot. *New Scientist*, vol. 176, 30–33.

Grandin, T. (2006). *Thinking in pictures, expanded edition: my life with autism.* New York, Vintage Books.

Green, E. & A. Adam. (1998). On-line leisure: gender, and ICTs in the home. *Information Communication & Society*, vol. 1, no. 3, 291–312.

Habermas, J. (1970). *Towards a rational society.* Boston, Beacon Press.

(1984). *The theory of communicative action: lifeworld and system: a critique of functionalist reason.* Boston, Beacon.

Hacking, I. (1983). *Representing and intervening: introductory topics in the philosophy of natural science.* Cambridge & New York, Cambridge University Press.

Hansen, A.H. (1921). The technological interpretation of history. *Quarterly Journal of Economics*, vol. 36. no 1, November, 72–83.

Happé, F. (2000). Parts and wholes, meaning and minds: central coherence and its relation to theory of mind. In *understanding other minds: perspectives from developmental cognitive neuroscience*, eds. S. Baron-Cohen, H. Tager-Flusberg & D.J. Cohen, 203–221. Oxford, Oxford University Press.

Haraway, D. (1985). A manifesto for cyborgs – science, technology, and socialist feminism in the 1980s. *Socialist Review*, no. 80, 65–107.

Harding, S.G. (1986). *The science question in feminism.* Ithaca, NY, Cornell University Press.

Harman, G. (2009). *Prince of networks: Bruno Latour and metaphysics.* Melbourne, Australia, Re.press.

(2011). *The quadruple object.* Windchester, UK, Zero Books.

(2013). An outline of object-oriented philosophy. *Science Progress*, vol. 96, no. 2, 187–199.

Harré, R. & E. Madden. (1975). *Causal powers.* Oxford, Basil Blackwell.

Harvey, D. (1989). *The condition of postmodernity: an enquiry into the origins of cultural change.* Oxford, UK; Cambridge, MA, Blackwell.

(2006). *The limits to capital.* London; New York, Verso.

Hayles, K. (1999). *How we became posthuman: virtual bodies in cybernetics, literature, and informatics.* Chicago, IL, University of Chicago Press.

Heersmink, R. (2012). Defending extension theory: A response to Kiran and Verbeek. *Philosophy & Technology*, vol. 25, no. 1, 121–128.

Heft, H. (1989). Affordances and the body: an intentional analysis of Gibson's ecological approach to visual perception. *Journal for the Theory of Social Behaviour*, vol. 19, no. 1, 1–30.

Heidegger, M. (1962). *Being and time.* Malden, MA, Blackwell (1978).

(1977). *The question concerning technology, and other essays.* New York, Harper & Row.

Heilbroner, R.L. (1967). Do machines make history? *Technology and Culture*, vol. 8, 335–345.

Hennessy, J.R. (2014) 22 July. The tech utopia nobody wants: why the world nerds are creating will be awful. *The Guardian*. www.theguardian.com /commentisfree/2014/jul/22/the-tech-utopia-nobody-wants-why-the-wor ld-nerds-are-creating-will-be-awful.

Hermelin, B. (1970). *Psychological experiments with autistic children, by B. Hermelin and N. O'Connor.* Oxford, New York, Pergamon Press.

Higgs, E., A. Light & D. Strong. (2000). *Technology and the good life?* Chicago, IL, University of Chicago Press.

Hobson, R.P. (1993). The emotional origins of social understanding. *Philosophical Psychology*, vol. 6, no. 3, 227–249.

Houkes, W., P. Kroes, A. Meijers & P.E. Vermaas. (2011). Dual-nature and collectivist frameworks for technical artefacts: a constructive comparison. *Studies in History and Philosophy of Science Part A*, vol. 42, no. 1, 198–205.

Houkes, W. & P.E. Vermaas. (2010). *Technical functions: on the use and design of artefacts.* Dordrecht, Springer.

Hughes, T. (1983). *Networks of power: electrification in Western society, 1880–1930.* Baltimore and London, John Hopkins University Press.

Hume, D., T.H. Green & T.H. Grose. (2001). *Of commerce.* University of Virginia Library.

Hutchins, E. (1995). *Cognition in the wild.* Cambridge, MA, MIT Press.

Ihde, D. (1983). The historical-ontological priority of technology over science. In *Philosophy and Technology*, eds. P. Durbin & F. Rapp, 235–252. Netherlands, Springer.

(1990). *Technology and the lifeworld: from garden to earth.* Bloomington, Indiana University Press.

(1991). *Instrumental realism: the interface between philosophy of science and philosophy of technology.* Bloomington, Indiana University Press.

(1998). *Expanding hermeneutics: visualism in science.* Evanston. Northwestern University Press.

(2009). Technology and science. In *A companion to the philosophy of technology*, eds. J.-K.B. Olsen, S.A. Pedersen & V.F. Hendricks, 49–60. Chichester, UK & Malden, MA, Wiley-Blackwell.

Illych, I. (1973). *Tools for conviviality.* New York, Harper and Row.

Ingold, T. (2000). *The perception of the environment: essays in livelihood, dwelling and skill.* London & New York, Routledge.

Introna, L.D. (2009). Ethics and the speaking of things. *Theory, Culture & Society*, vol. 26, no. 4, 25–46.

Jessop, B. (2009). The spatiotemporal dynamics of globalizing capital and their impact on state power and democracy. In *High-speed society: social acceleration, power, and modernity*, eds. H. Rosa & W.E. Scheuerman, 135–160. University Park, PA, Pennsylvania State University Press.

Johnson, D. (2010). Sorting out the questions of feminist technology. In *Feminist technology: women, gender, and technology*, eds. L.L. Layne, S.L. Vostral & K. Boyer, 197–202. Urbana, University of Illinois Press.

Jones, M.R. & H. Karsten. (2008). Giddens's structuration theory and information systems research. *MIS Quarterly*, vol. 32, no. 1, 127–157.

Kallinikos, J., A. Aaltonen & A. Marton. (2010). A theory of digital objects. First Monday. http://journals.uic.edu/ojs/index.php/fm/article/view/30 33/2564

Kanner, L. (1943). Autistic disturbances of affective contact. *Nervous Child*, vol. 2, 217–250.

Kant, I. (1784). Idea for a universal history from a cosmopolitan point of view. www.marxists.org/reference/subject/ethics/kant/universal-history.htm.

Kant, I. & N.K. Smith. (2003). *Critique of pure reason*. Houndmills, Basingstoke, Hampshire & New York, Palgrave Macmillan.

Kaplan, A. (1964). *The conduct of inquiry: methodology for behavioural science*. San Francisco, Chandler Publishing.

Kaplan, D.M. (2009). *Readings in the philosophy of technology*. Lanham, Rowman & Littlefield Publishers.

Kapp, E. (1877). *Grundlinien einer Philosophie der Tecknik*. Braunmschwieg, Germany, Westermann.

Kiran, A. & P.-P. Verbeek. (2010). Trusting ourselves to technology. *Knowledge, Technology & Policy*, vol. 23, no. 3, 409–427.

Kirkpatrick, G. (2008). *Technology and social power*. New York, Palgrave Macmillan.

Kirkup, G. (2000). *The gendered cyborg: a reader*. London & New York, Routledge in association with the Open University.

Kline, R. (1995). Construing technology as applied science – public rhetoric of scientists and engineers in the United-States, 1880–1945. *Isis*, vol. 86, no. 2, 194–221.

Kling, R. (1991). Computerization and social transformations. *Science, Technology & Human Values*, vol. 16, no. 3, 342–367.

Koffka, C. (1935). *Principles of Gestalt psychology*. Kegan Paul & Co.: London.

Koselleck, R. (1985). *Futures past: on the semantics of historical time*. Cambridge, MA, MIT Press.

(2004). *Futures past: on the semantics of historical time*. New York & Chichester, Columbia University Press.

Kraybill, D.B., K.M. Johnson-Weiner & S.M. Nolt. (2013). *The Amish*. JHU Press.

Kroeber, A.L. (1948). *Anthropology: race. language, culture, psychology, pre-history*. New York, Harcourt.

Kroes, P. (2010). Engineering and the dual nature of technical artefacts. *Cambridge Journal of Economics*, vol. 34, no. 1, 51–62.

(2012). *Technical artefacts: creations of mind and matter a philosophy of engineering design*. Dordrecht, Springer.

Kroes, P. & A. Meijers. (2006). The dual nature of technical artefacts – introduction. *Studies in History and Philosophy of Science*, vol. 37, no. 1, 1–4.

Kroes, P.A. & A.W.M. Meijers. (2000). Introduction. In *The empirical turn in the philosophy of technology*, eds. P.A. Kroes & A.W.M. Meijers. Amsterdam, JAI/ Elsevier.

Kusch, M. (1997). The sociophilosophy of folk psychology. *Studies in History and Philosophy of Science Part A*, vol. 28, no. 1, 1–25.

Laclau, E. & C. Mouffe. (2001). *Hegemony and socialist strategy: towards a radical democratic politics*. London & New York, Verso.

Latour, B. (1987). *Science in action: how to follow scientists and engineers through society*. Cambridge, MA, Harvard University Press.

(1988). *The pasteurization of France*. Cambridge, MA, Harvard University Press.

(1993). *We have never been modern*. New York; London, Harvester Wheatsheaf.

(1994). Where are the missing masses? The sociology of a few mundane artifacts. In *Shaping technology/building society: studies in sociotechnical change*, eds. W.E. Bijker & J. Law, 225–258. Cambridge, MA, MIT Press.

(1999). For David Bloor ... and beyond: A reply to David Bloor's' anti-Latour. *Studies in history and philosophy of science*, vol. 30, 113–130.

(2004). *Politics of nature: how to bring the sciences into democracy*. Cambridge, MA, Harvard University Press.

(2005). *Reassembling the social: an introduction to actor network theory*. Oxford, Oxford University Press.

(2009). *The science of passionate interests: an introduction to Gabriel Tarde's economic anthropology*. Chicago, IL, Prickly Paradigm Press.

(2015). *An inquiry into modes of existence: an anthropology of the moderns*. Cambridge, MA, Harvard University Press.

Latsis, J., C. Lawson & N. Martins. (2007). *Contributions to social ontology*. London, Routledge.

Laudan, R. (1984). *Cognitive change in technology and science*. Netherlands, Springer.

Law, J. (2004). *After method: mess in social science research*. London, Routledge.

Lawson, C. (1996). Holism and collectivism in the work of J.R. Commons. *Journal of Economic Issues*, vol. 30, no. 4, 1–18.

(2007). Technology, technological determinism and the transformational model of technical activity. In *Contributions to social ontology*, eds. C. Lawson, J. Latsis & N. Martins, 32–49. London, Routledge.

(2008). An ontology of technology: artefacts, relations and functions. *Techné: Research in Philosophy and Technology*, vol. 12, no. 1. 48–64.

(2009). Ayres, technology and technical objects. *Journal of Economic Issues*, vol. 43, no. 3, 641–659.

(2010). Technology and the extension of human capabilities. *Journal for the Theory of Social Behaviour*, vol. 40, no. 2, 207–223.

(2017). Feenberg, Rationality and Isolation. In *Theory and Practice – Critical Theory and the Thought of Andrew Feenberg*, eds. D. Arnold & M. Andreas. Basingstoke, Palgrave MacMillan.

Lawson, J. (2003). Depth accessibility difficulties: An alternative conceptualisation of autism spectrum conditions. *Journal for the Theory of Social Behaviour*, vol. 33, no. 2, 189–202.

(2013). Economics and autism: why the drive towards closure? In *Contributions to social ontology*, eds. C. Lawson, J.S. Latsis & N. Martins. Abingdon, Routledge.

Lawson, J., S. Baron-Cohen & S. Wheelwright. (2004). Empathising and systemising in adults with and without Asperger syndrome. *Journal of Autism and Developmental Disorders*, vol. 34, no. 3, 301–310.

Lawson, T. (1992). Abstraction, tendencies and stylized facts. In *Real-life economics: Understanding wealth creation*, eds. P. Ekins & M. Max-Neef. London & New York, Routledge.

(1997). *Economics and Reality*. London, Routledge.

(2012). Ontology and the study of social reality: emergence, organisation, community, power, social relations, corporations, artefacts and money. *Cambridge Journal of Economics*, vol. 36, no. 2, 345–385.

(2013). Emergence and morphogenesis: causal reduction and downward causation? In *Social morphogenesis*, ed. M.S. Archer, 61–84. Netherlands, Springer.

(2015). A conception of social ontology. In *Social Ontology and Modern Economics*, ed. S. Pratten, 19–52. Abingdon, Oxford, Routledge.

(2016). Social positioning and the nature of money. *Cambridge Journal of Economics*, vol. Forthcoming.

Layne, L.L., S.L. Vostral & K. Boyer. (2010). *Feminist technology*. Urbana, University of Illinois Press.

Layton, E.T. (1974). Technology as knowledge. *Technology and culture*, 31–41.

Leslie, A.M. (1987). Pretense and representation – the origins of theory of mind. *Psychological Review*, vol. 94, no. 4, 412–426.

Libby, S., S. Powell, D. Messer & R. Jordan. (1998). Spontaneous play in children with autism: a reappraisal. *Journal of Autism and Developmental Disorders*, vol. 28, no. 6, 487–497.

Lohan, M. & W. Faulkner. (2004). Masculinities and technologies. *Men and Masculinities*, vol. 6, no. 4, 319–329.

Lord, C., D.J. Merrin, L.O. Vest & K.M. Kelly. (1983). Communicative behavior of adults with an autistic 4-year-old boy and his nonhandicapped twin brother. *Journal of Autism and Developmental Disorders*, vol. 13, no. 1, 1–17.

Lotka, A. (1956). *Elements of mathematical biology*. New York, Dover.

Loveland, K.A. (1991). Social affordances and interaction II: autism and the affordances of the human environment. *Ecological Psychology*, vol. 3, no. 2, 99–119.

Loveland, K.A., R.E. Mcevoy, B. Tunali & M.L. Kelley. (1990). Narrative story telling in autism and Down's syndrome. *British Journal of Developmental Psychology*, vol. 8, 9–23.

Loveland, K.A. & B. Tunali. (1991). Social scripts for conversational interactions in autism and down-syndrome. *Journal of Autism and Developmental Disorders*, vol. 21, no. 2, 177–186.

Lubbe, H. (1994). Keeping pace with time, on curtailed presence in the present. *Voprosy Filosofii*, no. 4, 94–107.

(2009). The contraction of the present. In *High-speed society: social acceleration, power, and modernity*, eds. H. Rosa & W.E. Scheuerman, vi, 313. University Park, PA, Pennsylvania State University Press.

Lukács, G. (1971). *History and class consciousness; studies in Marxist dialectics*. Cambridge, MA, MIT Press.

Lykke, N. (2010). *Feminist studies: a guide to intersectional theory, methodology and writing*. New York, Routledge.

Lynch, M. (1992). Going full circle in the sociology of knowledge: comment on Lynch and Furman. *Science, Technology and Human Values*, vol. 17, 228–233.

Mackenzie, D. (1984). Marx and the machine. *Technology and Culture*, vol. 25, no. 3, 473–502.

MacKenzie, D.A. & J. Wajcman. (1985a). Introduction. *In The social shaping of technology*, eds. D.A. MacKenzie & J. Wajcman, 2–27. Buckingham, Philadelphia, Open University Press.

(1985b). *The social shaping of technology*. eds. D.A. MacKenzie & J. Wajcman. Buckingham, Philadelphia, Open University Press.

Mandel, E. (1975). *Late capitalism*. London, Humanities Press.

Manovich, L. (2002). *The language of new media*. Cambridge, MA, MIT Press.

(2013). *Software takes command*. New York; London, Bloomsbury.

Marcuse, H. (1964). *One-dimensional man*. Boston, Beacon Press.

Margolis, E. & S. Laurence. (2007). *Creations of the mind: theories of artifacts and their representation.* Oxford & New York, Oxford University Press.

Markus, M.L. & D. Robey. (1988). Information technology and organizational change: causal structure in theory and research. *Management Science,* vol. 34, no. 5, 583–598.

Marx, K. (1955 [1900]). *The poverty of philosophy.* Moscow, Progress Publishers.

(1956). *The poverty of philosophy.* Digireads. com Publishing.

(1972 [1859]). *A contribution to the critique of political economy.* London, Lawrence and Wishart.

Marx, L. (2010). Technology: The emergence of a hazardous concept. *Technology and Culture,* vol. 51, no. 3, 561–577.

Marx, L. & B. Mazlish. (1996). *Progress: fact or illusion?* Ann Arbor, University of Michigan Press.

Marx, L. & M.R. Smith. (1994). *Does technology drive history?: the dilemma of technological determinism.* Cambridge, MA, MIT Press.

Max, M.L. & J.C. Burke. (1997). Virtual reality for autism communication and education, with lessons for medical training simulators. *Studies in Health Technology and Informatics,* vol. 39, 46–53.

McArthur, L.Z. & R.M. Baron. (1983). Toward an ecological theory of social perception. *Psychological Review,* vol. 90, 215–238.

McLaughlin, P. (2001). *What functions explain: functional explanation and self-reproducing systems.* Cambridge, Cambridge University Press.

McLuhan, M. (1964). *Understanding media: the extensions of man.* New York, McGraw-Hill.

Meijers, A. (2009). Philosophy of technology and engineering sciences. General Introduction. *Philosophy of Technology and Engineering Sciences,* vol. 9, 1–19.

Merchant, C. (1989). *The death of nature: women, ecology, and the scientific revolution.* New York, Harper & Row.

Metcalfe, J.S. (2010). Technology and economic theory. *Cambridge Journal of Economics,* vol. 34, no. 1, 153–171.

Mill, J.S. (1840). M. de Toqueville on Democracy in America. *Edinburg Review,* vol. 2, October.

Misa, T. (2009). Defining technology. In *A companion to the philosophy of technology,* eds. J.-K.B. Olsen, S.A. Pedersen & V.F. Hendricks, 8–12. Chichester, UK & Malden, MA, Wiley-Blackwell.

Mitcham, C. (1979). Philosophy and the history of technology. In *The history and philosophy of technology,* eds. G. Bugliarello & D.B. Doner. Urbana, IL & London, University of Illinois Press.

(1990). Three ways of being-with technology. In *From artifact to habitat: studies in the critical engagement of technology, Research in technology studies*, ed. G.L. Ormiston, 221. Bethlehem & London, Lehigh University Press & Associated University Presses.

(1994). *Thinking through technology: the path between engineering and philosophy*. Chicago, IL & London, University of Chicago Press.

Mokyr, J. (2000). Knowledge, technology, and economic growth during the industrial revolution. In *Productivity, technology and economic growth*, 253–292. New York, Springer.

(2002). *The gifts of Athena: historical origins of the knowledge economy*. Princeton, NJ, Princeton University Press.

Moore, M. & S. Calvert. (2000). Brief report: vocabulary acquisition for children with autism: teacher or computer instruction. *Journal of Autism and Developmental Disorders*, vol. 30, no. 4, 359–62.

Mundy, P., M. Sigman, J. Ungerer & T. Sherman. (1986). Defining the social deficits of autism: the contribution of non-verbal communication measures. *Journal of Child Psychology and Psychiatry and Allied Disciplines*, vol. 27, no. 5, 657–669.

Munkirs, J.R. (1988). The dichotomy: views of a fifth generation institutionalist. *Journal of Economic Issues*, vol. 22, no. 4, 1035–1044.

Murray, S. (2011). *Autism*. London, Routledge.

Neuhaus, E., T.P. Beauchaine & R. Bernier. (2010). Neurobiological correlates of social functioning in autism. *Clinical psychology review*, vol. 30, no. 6, 733–748.

Newman, M. G. (2004). *Technology in psychotherapy: An introduction*. *Journal of Clinical Psychology*, vol. 60, 141–145.

Nichols, S. & S. Stich. (2000). A cognitive theory of pretense. *Cognition*, vol. 74, no. 2, 115–147.

Noble, D.F. (1984). *Forces of production: a social history of industrial automation*. New York, Knopf.

Noys, B. (2014). *Malign velocities: accelerationism and capitalism*. John Hunt Publishing.

O'Loughlin, C. & P. Thagard. (2000). Autism and coherence: A computational model. *Mind & Language*, vol. 15, no. 4, 375–392.

Ogburn, W.F. (1922). *Social change with respect to culture and original nature*. New York, B.W. Huebsch, Inc.

(1933). *Living with machines*. Chicago, IL, American Library Association.

(1938). *Machines and tomorrow's world*. New York, Public Affairs Committee, Inc.

Oldenziel, R. (1999). *Making technology masculine: men, women and modern machines in America, 1870–1945*. Amsterdam, Amsterdam University Press.

Orlikowski, W.J. (1992). The duality of technology: rethinking the concept of technology in organizations. *Organization Science*, vol. 3, no. 3, 398–427.

(2007). Sociomaterial practices: exploring technology at work. *Organization Studies*, vol. 28, no. 9, 1435–1448.

(2010). The sociomateriality of organisational life: considering technology in management research. *Cambridge Journal of Economics*, vol. 34, no. 1, 125–141.

Pels, D. (1996). The Politics of symmetry. *Social Studies of Science*, vol. 26, 277–304.

Perner, J. & B. Lang. (2000). Theory of mind and executive function: is there a developmental relationship? In *Understanding other minds: perspectives from developmental cognitive neuroscience*, eds. S. Baron-Cohen, H. Tager-Flusberg & D. J. Cohen, 150–181. New York, Oxford University Press.

Perner, J., B. Lang & D. Kloo. (2002). Theory of mind and self-control: more than a common problem of inhibition. *Child Development*, vol. 73, no. 3, 752–767.

Pickering, A. (1995). *The Mangle of practice: time agency and science*. Chicago, IL, University of Chicago Press.

Pinch, T. & W. Bijker. (1987). The social construction of facts and artifacts: or how the sociology of science and the sociology of technology might benefit from each other. In *The social construction of technological systems: new directions in the sociology and history of technology*, eds. T. Pinch, W. Bijker & T. Hughes. Cambridge, MA, MIT Press.

Pitt, J.C. (2000). *Thinking about technology*. New York, Seven Bridges Press.

Plant, S. (1997). *Zeroes + ones: digital women + the new technoculture*. New York, Doubleday.

Pleasants, N. (1996). Nothing is concealed: de-centring tacit knowledge and rules from social theory. *Journal for the Theory of Social Behaviour*, vol. 26, no. 3, 233–255.

Polanyi, K. (1957). *The great transformation: the political and economic origins of our time*. Boston, Beacon Press.

Porpora, D.V. (1989). Four concepts of social structure. *Journal for the Theory of Social Behaviour*, vol. 19, no. 2, 195–211.

Postman, N. (1993). *Technopoly: the surrender of culture to technology*. New York, Vintage Books.

Pratten, S. (2013). Critical realism and the process account of emergence. *Journal for the Theory of Social Behaviour*, vol. 43, no. 3, 251–279.

(2014). *Social ontology and modern economics*. Abingdon, Routledge.

(2015). *Social ontology and modern economics*. Abingdon, Routledge.

Preston, B. (1998). Why is a wing like a spoon? A pluralist theory of function. *The Journal of Philosophy*, vol. 95, no. 5, 215–254.

(2003). Of Marigold Beer – a reply to Vermaas and Houkes. *The British Journal for the Philosophy of Science*, vol. 54, 601–612.

Quine, W.V. (1953). *From a logical point of view: 9 logico-philosophical essays*. Cambridge, Harvard University Press.

Rathje, W.L. & M.B. Schiffer. (1982). *Archaeology*. San Diego, Harcourt, Brace and Jovanovich.

Reckwitz, A. (2002). Toward a theory of social practices: a development in culturalist theorizing. *European Journal of Social Theory*, vol. 5, no. 2, 243–263.

Reed, E. (1988). *James J. Gibson and the psychology of perception*. New Haven, Yale University Press.

Rifkin, J. (1987). *Time wars: the primary conflict in human history*. New York, H. Holt.

Robins, B., K. Dautenhahn, E. Ferrari, G. Kronreif, B. Prazak-Aram, P. Marti, I. Iacono, G.J. Gelderblom, T. Bernd & F. Caprino. (2012). Scenarios of robot-assisted play for children with cognitive and physical disabilities. *Interaction Studies*, vol. 13, no. 2, 189–234.

Robinson, J.P. & G. Godbey. (1999). *Time for life: the surprising ways Americans use their time*. University Park, PA, Pennsylvania State University Press.

Roelfsema, M.T., R.A. Hoekstra, C. Allison, S. Wheelwright, C. Brayne, F.E. Matthews & S. Baron-Cohen. (2012). Are autism spectrum conditions more prevalent in an information-technology region? A school-based study of three regions in the Netherlands. *J Autism Dev Disord*, vol. 42, no. 5, 734–739.

Rogers, S.J. & D.L. Dilalla. (1991). A comparative study of the effects of a developmentally based instructional model on young children with autism and young children with other disorders of behavior and development. *Topics in Early Childhood Special Education*, vol. 11, no. 2, 29–47.

Rosa, H. (2009). Social acceleration: ethical and political consequences of a desynchronized high-speed society. In *High-speed society: social acceleration, power, and modernity*, eds. H. Rosa & W.E. Scheuerman, 78–111. University Park, PA, Pennsylvania State University Press.

(2013). *Social acceleration: a new theory of modernity*. New York, Columbia University Press.

Rosa, H. & W.E. Scheuerman. (2009). *High-speed society: social acceleration, power, and modernity.* University Park, PA, Pennsylvania State University Press.

Rosenberg, N. (1976). Marx as a student of technology. *Monthly Review,* vol. 28, no. 3, 56–77.

Rothenberg, D. (1993). *Hand's end: technology and the limits of nature.* Berkeley, University of California Press.

Rousseau, J.-J. (1992). *Discourse on the sciences and arts: (first discourse) and polemics.* Hanover, Dartmouth College.

Rubinstein, D. (2013). *Marx and Wittgenstein: Social praxis and social explanation.* London, Routledge.

Rutherford, M. (2011). *The institutionalist movement in American economics, 1918–1947: science and social control.* Victoria, Cambridge University Press.

Rynkiewicz, A., B. Schuller, E. Marchi, S. Piana, A. Camurri, A. Lassalle & S. Baron-Cohen. (2016). An investigation of the 'female camouflage effect' in autism using a computerized ADOS-2 and a test of sex/gender differences. *Molecular Autism,* vol. 7, no. 1, 1–8.

Salomon, G. (1993). *Distributed cognitions: psychological and educational considerations.* Cambridge & New York, Cambridge University Press.

Sayer, A. (1997). Essentialism, social constructionism, and beyond. *Sociological Review,* vol. 45, no. 3, 453–487.

Scharff, R. & V. Dusek. (2003). *Philosophy of technology: the technological condition, an anthology.* London, Blackwell Publishing.

Schatzberg, E. (2006). Technik comes to America: changing meanings of technology before 1930. *Technology and Culture,* vol. 47, no. 3, 486–512.

Schatzki, T.R. (1996). *Social practices: a Wittgensteinian approach to human activity and the social.* New York, Cambridge University Press.

Schatzki, T.R., K. Knorr-Cetina & E.V. Savigny. (2001). *The practice turn in contemporary theory.* New York, Routledge.

Schiffer, M.B. (1992). *Technological perspectives on behavioural change.* Tucson, University of Arizona Press.

Schyfter, P. (2009). The bootstrapped artefact: a collectivist account of technological ontology, functions, and normativity. *Studies in History and Philosophy of Science Part A,* vol. 40, no. 1, 102–111.

Scott, S. (2001). Metarepresentation in philosophy and psychology. In *Proceedings of the 23rd Annual Conference of the Cognitive Science Society.* Human Communication Research Centre, University of Edinburgh, 910–916.

Searle, J.R. (1995). *The construction of social reality.* London, Penguin Books.

(2010). *Making the social world: the structure of human civilization.* Oxford; New York, Oxford University Press.

Secord, P. (1986). Explanation in the social sciences and in life situations. In *Metatheory in social science*, eds. D.W. Fiske & R.A. Schwartz. Chicago, IL, University of Chicago Press.

Shalom, D.B. (2009). The medial prefrontal cortex and integration in autism. *The Neuroscientist*, vol. 15, no. 6, 589–598.

Shapin, S. (1982). History of science and its sociological reconstructions. *History of Science*, vol. 20, no. 1982, 157–211.

Shaw, W.H. (1979). 'The handmill gives you the feudal lord': Marx's technological determinism. *History and Theory*, vol. 18, no. 2, 155–176.

Shove, E., M. Pantzar & M. Watson. (2012). *The dynamics of social practice: everyday life and how it changes.* Los Angeles, Sage.

Sigman, M. & P. Mundy. (1989). Social attachments in autistic children. *Journal of the American Academy of Child and Adolescent Psychiatry*, vol. 28, no. 1, 74–81.

Sigman, M., J.A. Ungerer & A. Russell. (1983). Moral judgment in relation to behavioral and cognitive disorders in adolescents. *Journal of Abnormal Child Psychology*, vol. 11, no. 4, 503–511.

Sigman, M.D., C. Kasari, J.H. Kwon & N. Yirmiya. (1992). Responses to the negative emotions of others by autistic, mentally retarded, and normal children. *Child Development*, vol. 63, no. 4, 796–807.

Silberman, S. (2000). The Geek syndrome. Wired. www.wired.com/2001/12/ aspergers/.

(2015). *Neurotribes: the legacy of autism and the future of neurodiversity.* New York, Penguin.

Simmel, G. (1991). Money in modern culture. *Theory Culture & Society*, vol. 8, no. 3, 17–31.

(2004). *The philosophy of money.* London, Routledge.

Simon, H.A. (1996). *The sciences of the artificial.* Cambridge, MA, MIT Press.

Simondon, G. (1964). *L'individu et sa genèse physico-biologique; l'individuation à la lumière des notions de forme et d'information.* Paris, Presses universitaires de France.

(1980). *On the mode of existence of technical objects.* London, Ontario, University of Ontario Press.

Sismondo, S. (2004). *An introduction to science and technology studies.* Malden, MA, Blackwell Pub.

Skolimowski, H. (1966). The structure of thinking in technology. *Technology and Culture*, vol. 7, no. 3, 371–383.

Smil, V. (2005). *Creating the twentieth century: technical innovations of 1867–1914 and their lasting impact.* Oxford; New York, Oxford University Press.

Smith, B. (2003). Ontology. In *Blackwell guide to the philosophy of computing and information*, ed. L. Floridi, 155–166. Oxford, Blackwell.

Smith, B. & J. Searle. (2003). An illuminating exchange the construction of social reality. *American Journal of Economics and Sociology*, vol. 62, no. 1, 285–309.

Smith, D.L. & G.P. Ginsburg. (1989). The social perception process: reconsidering the role of social stimulation. *Journal for the Theory of Social Behaviour*, vol. 19, no. 1, 31–45.

Smith, I.M. & S.E. Bryson. (1994). Imitation and action in autism: a critical review. *Psychological Bulletin*, vol. 116, no. 2, 259–273.

Snyder, A.W. (1998). Breaking mindset. *Mind & Language*, vol. 13, no. 1, 1–10.

Solow, R.M. (1957). Technical change and the aggregate production function. *Review of Economics and Statistics*, vol. 39, no. 3, 312–320.

Soltanzadeh, S. (2015). Humanist and nonhumanist aspects of technologies as problem solving physical instruments. *Philosophy & Technology*, vol. 28, no. 1, 139–156.

(2016). Questioning two assumptions in the metaphysics of technological objects. *Philosophy & Technology*, vol. 29, no. 2, 127–135.

Stanley, A. (1995). *Mothers and daughters of invention: notes for a revised history of technology*. New Brunswick, NJ, Rutgers University Press.

Staudenmaier, J. (1995). Problematic stimulation: historians and sociologists constructing technology studies. In *Research in philosophy and technology, social and philosophical constructions of technology*, vol. 15, ed. M. Carl. Greenwich, Connecticut, JAI Press.

Steinert, S. (2015). Taking stock of extension theory of technology. *Philosophy & Technology*, vol. 29, no. 1, 61–78.

Stewart Millar, M. (1998). *Cracking the gender code: who rules the wired world?* Toronto, Second Story Press.

Stodgell, C.J., J.L. Ingram & S.L. Hyman. (2001). The role of candidate genes in unraveling the genetics of autism. *International Review of Research in Mental Retardation*, vol. 23, 57–81.

Strassmann, W.P. (1974). Technology: a culture trait, a logical category, or virtue itself. *Journal of Economic Issues*, vol. 8(4), 671–687.

Sturmey, P. (2003). Video technology and persons with autism and other developmental disabilities: an emerging technology for PBS. *Journal of Positive Behavior Interventions*, vol. 5, no. 1, 3–4.

Suchman, L. (2007). *Human-machine reconfigurations: plans and situated actions*. Cambridge, Cambridge University Press.

(2008). Feminist STS and the sciences of the artificial. In *The handbook of science and technology studies*, eds. E. Hackett, O. Amsterdamska, M. Lynch & J. Wacjman. Cambridge, MA, MIT Press.

Swallow Richards, E. (1911). The elevation of applied science to an equal rank with the so-called learned professions. In *Technology and industrial efficiency: a series of papers presented at the Congress of Technology*, 124–128. New York, McGraw Hill.

Taylor, B.A., C.E. Hughes, E. Richard, H. Hoch & A. Rodriquez Coello. (2004). Teaching teenagers with autism to seek assistance when lost. *Journal of Applied Behavior Analysis*, vol. 37, no. 1, 79–82.

Taylor, C. (1971). Interpretation and the sciences of man. *The review of metaphysics*, 3–51.

Turkle, S. (2005). *The second self: computers and the human spirit.* Cambridge, MA, MIT Press.

Uddin, L.Q. (2011). The self in autism: an emerging view from neuroimaging. *Neurocase*, vol. 17, no. 3, 201–208.

Ullman, E. (1997). *Close to the machine: technophilia and its discontents: a memoir.* San Francisco, City Lights Books.

Vaccari, A. (2013). Artifact dualism, materiality, and the hard problem of ontology: some critical remarks on the dual nature of technical artifacts program. *Philosophy & Technology*, vol. 26, no. 1, 7–29.

Valenti, S.S. & J.M.M. Gold. (1991). Social affordances and interaction: introduction. *Ecological Psychology*, vol. 3, no. 2, 77–98.

Vargas, D.L., C. Nascimbene, C. Krishnan, A.W. Zimmerman & C.A. Pardo. (2005). Neuroglial activation and neuroinflammation in the brain of patients with autism. *Annals of neurology*, vol. 57, no. 1, 67–81.

Veblen, T. (1908). On the nature of capital (part 1). *Quarterly Journal of Economics*, vol. 22, 517–542.

(1921). *The engineers and the price system.* New York, B. W. Huebsch.

Vermaas, P.E. & W. Houkes. (2006). Technical functions: A drawbridge between the intentional and structural natures of technical artefacts. *Studies in History and Philosophy of Science Part A*, vol. 37, no. 1, 5–18.

Vermeulen, P. (2012). *Autism as context blindness.* Shawnee Mission, Kan., AAPC Pub.

Vincenti, W.G. (1990). *What engineers know and how they know it.* Baltimore, Johns Hopkins University Press.

Virilio, P. (2006). *Speed and politics.* Los Angeles, CA, Semiotext(e).

Vostral, S.L. & D. McDonough. (2010). A feminist Inventor's Studio. In *Feminist technology: women, gender, and technology*, eds. L.L. Layne, S.L. Vostral & K. Boyer, 197–202. Urbana, University of Illinois Press.

Wainer, A.L. & B.R. Ingersoll. (2011). The use of innovative computer technology for teaching social communication to individuals with autism spectrum disorders. *Research in Autism Spectrum Disorders*, vol. 5, no. 1, 96–107.

Wajcman, J. (1991). *Feminism confronts technology*. University Park, PA, Pennsylvania State University Press.

(2004). *TechnoFeminism*. Cambridge; Malden, MA, Polity.

(2010). Feminist theories of technology. *Cambridge Journal of Economics*, vol. 34, no. 1, 143–152.

Waller, W.T. (1982). The evolution of the Veblenian dichotomy: Vablen, Hamilton, Ayres and Foster. *Journal of Economic Issues*, vol. 16, no. 3, 757–771.

Weber, M. (1958). *The Protestant work ethic and the spirit of capitalism*. New York, Scribners.

Weeks, S.J. & R.P. Hobson. (1987). The salience of facial expression for autistic children. *Journal of Child Psychology and Psychiatry and Allied Disciplines*, vol. 28, no. 1, 137–152.

Weitzman, M.L. (1996). Hybridizing growth theory. *American Economic Review*, vol. 86, no. 2, 207–212.

(1998). Recombinant growth. *Quarterly Journal of Economics*, vol. 113, no. 2, 331–360.

Wheelwright, S. & S. Baron-Cohen. (2001). The link between autism and skills such as engineering, maths, physics and computing: A reply to Jarrold and Routh, *Autism*, 1998, vol. 2, no. 3, 281–289. *Autism*, vol. 5, no. 2, 223–227.

Williams, E., A. Costall & V. Reddy. (1999). Children with autism experience problems with both objects and people. *Journal of Autism and Developmental Disorders*, vol. 29, no. 5, 367–378.

Williams, R. (1976). *Keywords: a vocabulary of culture and society*. London, Croom Helm.

Wing, L. (1974). *Children apart: autistic children and their families*. Washington, DC, National Society for Children with Adults with Autism.

Winner, L. (1978). *Autonomous technology. Technics-out-of-control as a theme in political thought*. Cambridge, MA, Massachusetts Institute of Technology Press.

(1980). Do artifacts have politics? *Daedalus*, vol. 109, no. 1, 121–136.

(1983). Technologies as forms of life. In *Epistemology, methodology, and the social sciences*, eds. R. Cohen & M. Wartofsky, 249–263. Netherlands, Springer.

(1991). Upon opening the black box and finding it empty: social constructivism and the philosophy of technology. In *The technology of discovery and discovery of technology*, eds. J. Pitt & E. Lugo. Blacksburg, VA, Society for the Philosophy of Technology.

Woolgar, S. (1988). *Science: the very idea.* Chichester, UK, Ellis Horwood.
(1991). The turn to technology in social studies of science. *Science, Technology & Human Values,* vol. 16, 20–50.

Von Wright, G.H. (1971). *Explanation and understanding.* London, Routledge & K. Paul.

Wulff, S.B. (1985). The symbolic and object play of children with autism: a review. *Journal of Autism and Developmental Disorders,* vol. 15, no. 2, 139–148.

Young, A.M.H., B. Chakrabarti, D. Roberts, M.-C. Lai, J. Suckling & S. Baron-Cohen. (2016). From molecules to neural morphology: understanding neuroinflammation in autism spectrum condition. *Molecular Autism,* vol. 7, no. 1, 1–8.

Zabel, R.H. & M.K. Zabel. (1982). Factors in burnout among teachers of exceptional children. *Exceptional Children,* vol. 49, no. 3, 261–263.

Index

Printed in the United States
By Bookmasters